Marine Life of the Caribbean

Alick Jones
Nancy Sefton

CARIBBEAN

First published 1979
Reprinted 1983, 1985, 1986, 1987, 1988, 1990 (twice), 1991

Published by MACMILLAN EDUCATION LTD
London and Basingstoke
Associated companies and representatives in Accra, Auckland, Delhi, Dublin, Gaborone, Hamburg, Harare, Hong Kong, Kuala Lumpur, Lagos, Manzini, Melbourne, Mexico City, Nairobi, New York, Singapore, Tokyo.

ISBN 0–333–25839–8

Printed in Hong Kong

Acknowledgements

The author and publishers wish to acknowledge, with thanks, the following photographic sources:

Bacon P. p 74 bottom
Biophotos p 68
Pritchard P. C. H. pp 74 top; 77 bottom; 78
Schultz J. p 77 top
Seddon S. A. pp 1; 4; 5; 21; 73
All other photographs courtesy of the authors

The publishers have made every effort to trace the copyright holders of all illustrations, but, where they have failed to do so they will be pleased to make the necessary arrangements at the first opportunity.

Contents

Dedication

Both geographically and culturally the Caribbean area is unique. However it shares with other regions and countries the challenge of controlling progress so that the natural environment does not suffer irreparable damage. The purpose of this book is to describe what Nature has given us, how we can enjoy it, and how we must protect it.

This book is thus dedicated to the children, in whose hands the future of the Caribbean lies. Let us hope that the horrific image conjured up by the following lines never becomes a reality:

> 'Once there were islands all a-sprout with palms: and coral reefs and sands as white as milk. What is there now but a vast shambles of the heart? Filth, squalor, and a world of little men.'
>
> *Titus Alone* Mervyn Peake

Acknowledgements

This book evolved (as is suitable for a Biology book) from the shorter 'Guide to the Marine Life of the Cayman Islands' published by the Cayman Islands Conservation Association in 1975. The original version was largely written by Nancy Sefton with help, guidance and support from many friends and colleagues. Alick Jones was approached by the Caribbean Conservation Association (C.C.A.) to see if it could be adapted into a Caribbean version. The text was extensively revised and expanded and new photographs were also added. This book is the result.

As well as thanking all those who helped with the earlier version we must also thank those whose aid came later. We particularly appreciate the contributions made by Dr. Peter Bacon of the University of the West Indies, Trinidad and Jill Sheppard, Executive Director of the C.C.A. The former reviewed and revised the sections on mangroves and turtles while the latter, as well as writing the section on the Caribbean and the Need for Conservation, was also a tower of organising strength from the C.C.A. Office. We can truly say that but for her we would neither of us have started this book let alone finished it. The C.C.A. and the University of the West Indies, Barbados, contributed funds to allow the two authors a period of valuable consultation and discussion in Cayman. Harriette Jordan typed almost all of the revised text.

Among the many who helped in other ways were the following: Jack Andresen, Angela Fields, Marco Giglioli, Lawrence Hull, Paul Jacques, Fran Miller, Ed Oliver, Andrea Rhodes, Glen Ullrich, John Warner, Steven K. Webster, Jim Wood, Jim and Cathy Church, Diversified Services Caribbean Ltd., Marticulture Ltd., M.R.C.U. (Mosquito Research Control Unit), Government of the Cayman Islands, Virgin Islands Conservation Society.

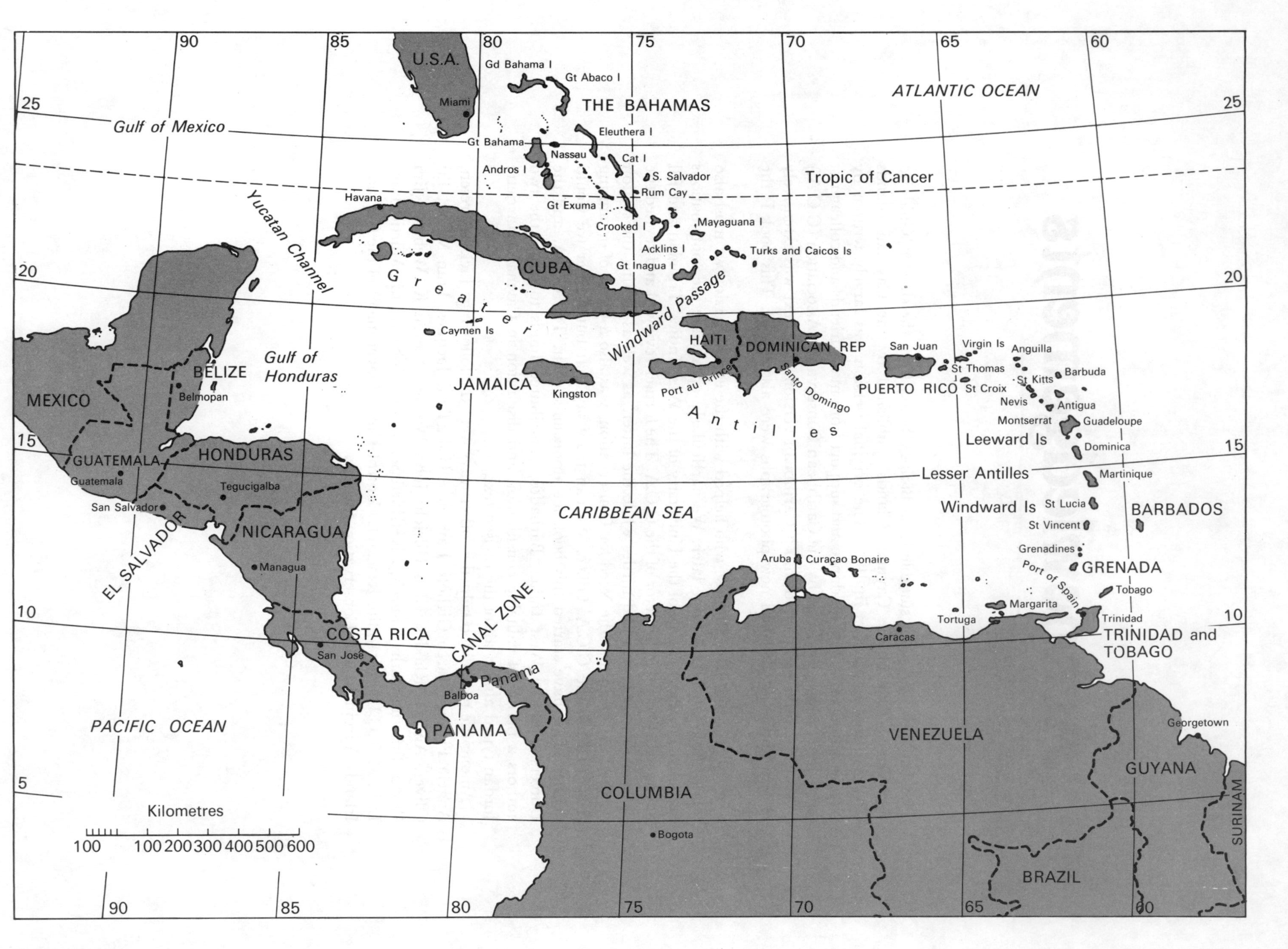

The Caribbean

The Caribbean and the need for Conservation

Modern living generates a great deal of material which does not readily decay. One of our major tasks is to ensure that as little as possible of this waste pollutes our environment.

The Caribbean means many different things to different people, whether they be inhabitants of the region or visitors.

The inhabitant will see the Caribbean from the point of view of his home. The Vincentian may think first of the Grenadines, St. Lucia and perhaps Barbados; the Guyanese, from the South American mainland, while very much aware of Venezuela and Brazil will feel that Trinidad is much nearer home; the Jamaican, while being closer, in geographical terms, to Cuba and Haiti than to any other Caribbean country, will still feel himself very much part of the English-speaking Caribbean. The Guadeloupean, while aware of the existence of Dominica because of its closeness and Martinique as his partner in the French Antilles, will tend to think firstly of France and only secondly of the Caribbean. A resident of Colombia, a Latin American country with a long coastline on the Caribbean, is unlikely to think of himself to any extent as a Caribbean person. The inhabitant of the Caribbean, in other words, is much more affected by his background and upbringing and by the political association of his homeland than by geographical factors.

The visitor, while to some extent influenced by local attitudes, will be likely to have an image of the region as a dream-world, a place of idyllic charm and rustic beauty, which is nevertheless conveniently equipped with many of the features of the sophisticated, developed, and to him, real world. This view is, of course, to a large extent due to the efforts of those in the Caribbean who, for economic reasons, are seeking to promote the region as a tourist paradise.

What is vital, mainly for its inhabitants but also for its visitors, is that the essential nature of the Caribbean should be preserved. In particular, we should be aware of the threats to the environment that development brings. Clearly a policy for conservation is necessary: but, as we have seen, the attitude of the 'Caribbean' varies from place to place. Understandably, therefore, views on conservation will also vary.

Some are concerned about the preservation of the cultural heritage. Indeed there is an urgent need to examine the whole historical background of the region, with its experiences, its failures and successes, especially those of its great men. Only in this way can the Caribbean become aware and proud of its identity. Thus attention must be paid to its historic buildings and sites, its museums and archives, and to its literature, music and art.

None of these activities will, however, be of any value unless attention is also paid to the natural heritage. In the countries of the Caribbean, especially the islands, natural resources are severely limited. While the mainland countries have thousands of square miles of agricultural land, forest, and often untapped mineral resources, an island such as Barbados or, Grand Cayman, has limited scope for agricultural production and must therefore make the most of the assets that it has in the shape of sun, sand, sea, clean air and a good water supply.

The problem faced by these small islands is one of preserving their natural features in the face of the threat to them posed by the demand for improved living standards; the building of more houses, the cutting down of trees, the making of roads, the establishment of factories and the need to dispose of their waste products into the air and sea. The encouragement of tourists also results in additional building and, in many cases, damage to the beaches, reefs and swamps.

It is the beaches, reefs and swamps, with all that they contain, that are the subject of this book. The Caribbean marine environment is one of immense richness, from the mangroves with their birds, insects and fish to the reefs with their corals of every imaginable size,

Pollution of our beaches by oil from ships many miles out to sea can only be controlled by international agreement and co-operation.

Less and less of our Caribbean coastline remains unspoiled. The region's developers must guard against destroying the physical beauty which is the basis of the Caribbean's attraction.

shape and colour, and their abundance of other animals and plants, to the deeper seas beyond the reefs with their characteristic fish.

It is an environment which is not only full of interest for the beholder but also of significance for the ecology and the economy of the whole region. Damage to any part of it whether by predatory visitors, sewage from hotels, effluents from factories, or oil discharged from tankers, sets up a chain reaction with far reaching effects not only on the fisheries and other aspects of the marine environment but also on the shape of the beaches, the nature of the land and the availability of food supplies. In time, very serious changes could take place which would not only destroy the natural beauty of these islands but also undermine the economic structure, particularly that part of it resulting from tourism.

It is hoped that the information contained in these pages will awaken in the reader, not only a desire for a closer examination of the treasures contained in the marine environment, but also a realisation of their beauty and a respect for the part they play in maintaining the integrity of the whole environment of the Caribbean.

The Caribbean: Gem of the Seas

Barbados. This is the Atlantic coast upon which the oceanic swells continually break. The leeward west coast is more sheltered.

The Caribbean Sea is an area of about 2.7 million square kilometres enclosed by the coasts of South and Central America, the large islands of the Greater Antilles and the smaller Lesser Antilles. For the purpose of this book we have also included the area to the north-east which contains the Bahamas. To the east the Caribbean is continuous with the Atlantic while to the north it merges into the Gulf of Mexico. This region, if one adds to it parts of the Gulf of Mexico and the Atlantic coast of Florida, constitutes a well defined biogeographical area. To the north the temperate eastern seaboard of the United States is very different from the tropics while to the south the great river systems of South America so affect the coastal regions that many of the marine animals and plants typical of the Caribbean are absent.

Over the central American isthmus is the Pacific Ocean with its related but different inhabitants.

The Land

Nine countries of continental America border the Caribbean. A considerable portion of this coastline is fringed with mangrove swamps thriving on the sediment washed down by the rivers and lining the sheltered bays and estuaries. There are many long sandy beaches sometimes with fine coral reefs offshore; those in Belize are particularly famous for their beauty.

The islands of the Caribbean vary in size from Cuba (114,524 square kilometres; larger than the state of Kentucky and twice the size of Ireland) down to the tiny uninhabited islets of the north-eastern part of the sea. They vary in geological origin and it is this that produces the great variety of coastal scenery. The large islands of the Greater Antilles: Cuba, Hispaniola, Jamaica and Puerto Rico, have a volcanic origin but active volcanism has ceased long ago. Sedimentary and coral rocks have been added to their igneous core and their long history has meant that there has been time for the rocks to be lifted, folded and eroded. As a result these islands have a complicated structure and many types of scenery.

The islands of the Lesser Antilles, stretching from the Virgin Islands in the north to Grenada in the south, are also of largely volcanic origin but they are much more recent with the youngest islands at the southern end of the chain. Volcanic activity is still widespread. In 1902, Mt. Pelée in Martinique erupted violently and it is said that all but one of the 40,000 inhabitants of St. Pierre perished. In the same year La Soufrière in

St. Vincent erupted killing 2,000 people. La Soufrière in Guadaloupe underwent violent activity in 1976 but there was no major eruption.

Hot springs abound on many of the islands and volcanically produced steam has a potential for generating electricity on at least two islands. The volcanic islands frequently have black beaches, the grains being composed of tiny larva rock fragments. Barbados is very much the exception to this general picture. It is a flat coral cap sitting on a core of much older sedimentary rock. It is the highest peak of a submarine ridge which is an extension of the mountain systems of South America. Antigua is not wholly volcanic either but has fairly recent limestone and other sedimentary rocks as well as igneous ones. Trinidad, the most southerly of the islands has only recently separated from Venezuela and like her possesses oil fields of great commercial value.

In one area of the Caribbean the islands lie in very shallow water. These are the Bahamas, made of continental rock to the north but rock of volcanic origin to the south. They are surrounded by tens of thousands of square kilometres of water only 10 metres or so deep. This shallow sea is a classic site of active limestone formation. The water is saturated with calcium carbonate and as the sun heats it up evaporation takes place resulting in the precipitation of the limestone. This process is thought to have been going on for 130 million years and the limestone created may be as much as 3,300 metres thick. As it is formed the whole mass sinks into the earth's crust because of its enormous weight.

The Sea

The Caribbean lies between about 27°N and 8°N mostly south of the Tropic of Cancer. As a result sea temperatures are high and fairly constant; for instance in Barbados they vary between 25.5°C and 27.7°C. Further north the sea is cooler and shows greater seasonal variation, but even in Florida the mean annual temperature is higher than 25°C and never falls below 20°C. This is particularly important as coral reef formation does not occur below the latter temperature.

The salinity is also high and stable. Only at the mouths of large rivers (eg. Orinoco) does fresh water affect the marine fauna to any extent. 'Bubbles' of water of reduced salinity from the mouth of the Amazon do travel northwards into the region, but by the time they arrive their influence is negligible.

Montserrat is a volcanic island with sharp pointed hills and active hot sulphur springs known as Soufrieres.

The major current movements in the Caribbean are an extension of the South Equatorial Current. This sweeps across the breadth of the Atlantic from the west coast of Africa, and along the north-east coast of South America. It passes into the Caribbean running in a roughly westerly direction (drift bottles released in Barbados have been recovered from Nicaragua). The current then turns to the north to pass into the Gulf of Mexico through the Yukatan Channel. This water finally passes out into the Atlantic again to form the Gulf Stream. Some surface water also enters the Caribbean from the North Equatorial Current passing both north and south of the Greater Antilles.

Much of the water moving in the South Equatorial Current has welled up from the deep ocean along the south-west coast of Africa in the Benguela Current. This cold uprising water is very rich in nutrients and the ocean in that region is very productive. However by the time the water has moved across the Atlantic the floating animals and plants have removed most of these nutrients. Many of these creatures are eaten or die and sink to the ocean floor. This means that the Caribbean waters are poor in nutrients, unproductive and contain relatively few living organisms; a situation typical of most tropical oceans and seas (see p. 28). Thus the fisheries of the region, although important and probably capable of some further exploitation, will never be as rich as those in many temperate regions or areas of upwelling.

However although the productivity of the open Caribbean is low this is not the case for some of the coastal areas. Both mangrove swamps and coral reefs are capable of producing anything up to 40 times as much living material per unit area as the barren sea. In the mangrove swamps the sparse nutrients of the sea are supplemented by drainage from the land as well as by large quantities of rotting leaves and so on from the trees and other plants. In coral reefs the high productivity is partly due to the way in which the nutrients are passed from one member of the community to another in an efficient way, so that they are prevented from leaving the reef to enter the open sea. The remarkable symbiosis between the coral organisms and their zooxanthellae (see p. 37) is also responsible for the striking difference between the richness of the reef and the poverty of the open sea.

Most of the Caribbean Sea is deep (the average depth is about 2,200 metres) so there is little area of continental shelf. This contrasts with the area around the large land masses which often have a fringing area many miles wide which is relatively shallow, averaging about 75 metres. Only in a few places are there large areas of the Caribbean which are so shallow (south of Cuba, the Bahamas Bank and between Jamaica and Nicaragua), elsewhere the land falls steeply into the sea. In one or two places the depths are great by any standards, for example the Cayman trench just south of the Cayman Islands is over 7,000 metres deep.

Steep shorelines combined with small tides, often only 20 centimetres or so, mean that the intertidal zone (that which is exposed to the air at low tide) is relatively narrow. This is another point of contrast between the Caribbean and many non-tropical areas.

The Weather

The weather in the Caribbean is usually stable and because of its position close to the equator seasonal influences are usually slight. The only feature of the weather to have a profound effect on the marine environment is the hurricane. These violent cyclonic storms are usually spawned over equatorial oceans, in our case the Atlantic. Once formed they move in a westerly direction until they reach the Caribbean where they often veer towards the north. Thus on the whole the more southerly islands suffer the least.

A hurricane is defined as a tropical storm with winds of over 75 mph (127 kmph) but not infrequently the velocities may rise well above this; perhaps to 150 mph (254 kmph) with gusts of over 200 mph (338 kmph).

There are about 10 hurricanes per year in or near this region and as well as the damage caused on land the sea waves they generate wreak havoc on the inshore environment, destroying coral reefs, eroding and damaging mangrove swamps and changing beach profiles. Such infrequent but catastrophic events may well be long term limiting factors in the establishment of viable reefs in what otherwise seem to be suitable sites. Hurricane Hattie which swept over part of the Belize coast in 1961 destroyed about 90 per cent. of the corals in some places and it has been established that recovery *at favourable sites* could take 25-30 years and much longer in other places.

Heavy rain can also cause heavy mortality of marine animals living in shallow water. In Kingston harbour, Jamaica, for example, almost complete destruction of attached marine animals was reported following rainstorms during which over 10 centimetres of rain fell in 24 hours. The harbour is, however, a very confined water mass and on open coasts rain water usually has little effect.

In some places fresh-water springs are to be found on the sea bed. The areas around them are usually much affected and only contain animals and plants that can tolerate this dilution of the salt water.

The Tides

Moon generated tides are usually small in the Caribbean, often only 20 centimetres or so. However it is not uncommon to hear the term 'tides' used (incorrectly) to describe ocean currents. The small lunar tides mean that there is only a small area of the inshore environment which is uncovered and covered twice daily. Indeed the wave and surf action often almost obliterates the tidal variation. In mangrove swamps however where waves are usually absent the tides have a real influence and many of the animals are adapted to cope with periods of both immersion and emersion.

The Shores

There are many types of shore to be found in the Caribbean. These will be dealt with here only briefly since each type is covered in more detail in a later chapter.

Mangrove swamps

In places where there is little surf action and a gently sloping sea bed mangrove swamps tend to form. These conditions are to be found in sheltered bays and river estuaries. In the latter case the environment may be profoundly affected by both the fresh water and the silt of the river. To many people these swamps are unappealing and are often regarded as areas which can be drained or filled and exploited in other ways. In reality they are a complex and fascinating world which, if used properly, have a great potential for food production.

Sandy shores

In many places there is too much surf for mangroves to grow and sand will accumulate to produce the familiar golden, white or black beaches. Offshore from such beaches there may or may not be a coral reef. Although they appear barren these beaches have their own special plants and animals adapted to living in shifting unstable surroundings.

Tidal variation in the West Indies is usually small. Here the sea is at low tide. At high tide it will reach the pale band on the rock near the top of the one metre ruler.

The trade winds and oceanic swells often produce spectacular wave-swept coasts on the north and eastern sides of the islands.

Sea-grass beds

In some sandy bays, especially those behind reefs that contain coral fragments mixed with the sand, one frequently finds beds of the marine flowering plants known generally as sea-grasses. These plants have extensive root-like rhizomes that bind together the sand and the coral fragments to make it much more stable. Because of the more stable substrate and the presence of the grasses themselves this habitat contains a great variety of creatures. Being shallow and quiet they are ideal for the novice snorkeler.

Rocky beaches

These exist in many forms from gently washed boulders to surf-pounded cliffs. As with sandy beaches there may or may not be an offshore coral reef. The hard rocky surfaces provide homes for a vast array of animals and plants, many of which live actually fastened to the rock. Here also one finds essentially marine animals living more or less out of the sea, only wetted by the spray.

Coral reefs

True coral cannot survive out of water (the so-called fire corals can live for a short time poking clear of the surface). Consequently, coral reefs are all submarine features. Undoubtedly they are the most complex, fascinating and, in many ways, important of the inshore communities. Many books and research papers have attempted to categorise coral reefs into various types, but this is not an easy task. Firstly, the experts are themselves not in agreement about the ways in which the various types should be defined. Secondly, the reefs are so variable and complex that they do not lend themselves to a neat system of pigeon-holing. Attempts to generalise are therefore fraught with difficulties and lay the writer open to numerous criticisms based on exceptions to his 'rules' and 'definitions'.

Nonetheless it may be helpful to the reader to mention two or three generally recognised types of reef always remembering that a given reef may not readily fit any of the descriptions:

Fringing reefs are found near the shore. They may start a few feet from the edge of the water but as they grow they sometimes expand in a seaward direction leaving a shallow lagoon behind.

Patch reefs may develop in water of medium depth on otherwise sandy sea beds.

Barrier reefs are formed some distance from the shore, from which they are separated by fairly deep water, almost too deep to be termed a lagoon. There is some dispute about the use of this term for Caribbean reefs and certainly they are a very different phenomenon to the Great Barrier Reef of Australia.

The Middle World of the Mangroves

Roots of the red mangrove resemble knobbly knees as they 'walk' into the water.

Fruits of the mangrove trees. The white mangrove (left), the black mangrove (centre) and the red mangrove (right).

Nearly 25 per cent. of the coastline of the Caribbean Sea is occupied by a rich natural community called the mangrove. It is dominated by trees of *Rhizophora mangle*, the red mangrove, and reaches its best development in estuaries, on sheltered coasts near the mouths of large rivers which bring down silt and on bordering coastal lagoons. Such conditions are found along the south coasts of Cuba, Puerto Rico and Jamaica, where the mangroves grow also on coral reefs, in the North Sound on Grand Cayman, the Gulf of Paria coast of Trinidad, areas of Central America such as Yucatan and Panama and northern South America in localities like the Cienaga Grande of Colombia. Small patches of mangrove vegetation occur on almost all the West Indian islands, but only in estuarine or lagoonal situations do mangrove swamps develop.

Rhizophora has a branching prop-root system which supports it on soft muds in the inter-tidal region. Once established, the trees encroach on the sea by further prop-root extension and have been described as 'walking out to sea'. The tangled mass of roots traps debris and silt and thus encourages sedimentation, so that the swamp is extended gradually seawards. This is further aided by the seedlings which grow on the seaward side of the *Rhizophora* as they fall from the parent tree. Unlike most familiar trees, the seed grows while the fruit is still attached to the branches. Soon a slender seedling emerges from the end of the fruit growing ever downwards. By the time it drops it is from 18 to 36 centimetres long, and dart-shaped. When it falls it penetrates some distance into the mud where it becomes established. The seedling grows rapidly, its height increasing to 60 centimetres in the first year.

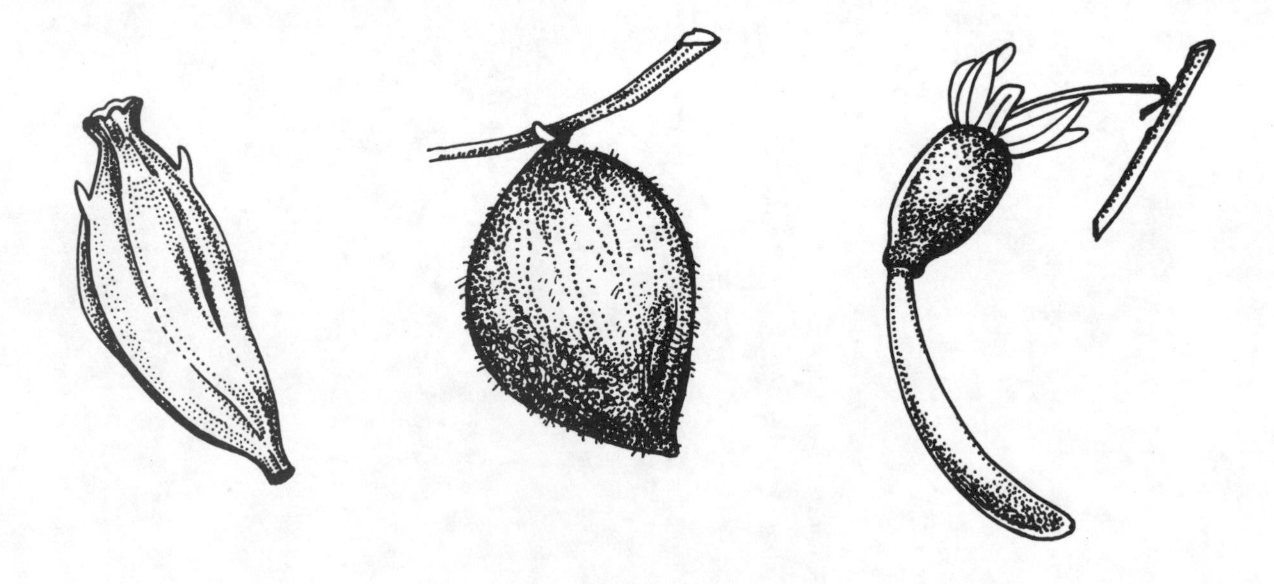

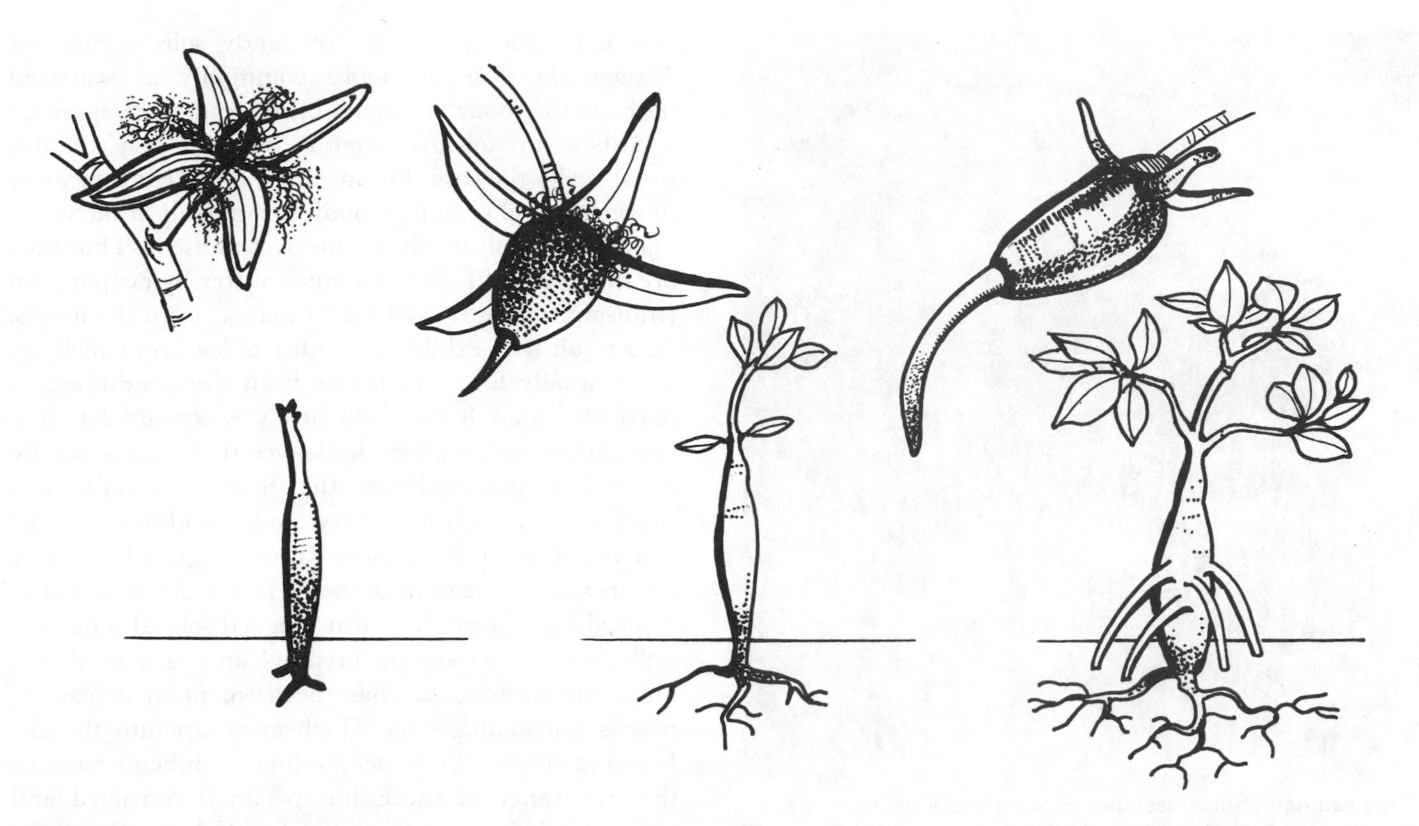

Stages in the development of the red mangrove tree.

In its second year, it begins to send out prop-roots for support. In this manner, a new mangrove area is born, or an old one expanded.

The red mangroves are 'pioneer' plants on coastal mudflats, but once the sediments are stabilised and their level raised to near high tide level, seeds of the black mangrove, *Avicennia germinans*, can germinate successfully. Lacking prop-roots, the black mangroves send small, woody fingers up through the ground. These pneumatophores enable the trees to obtain directly from the air the oxygen which is lacking in the peaty sediments. This dense root network produced by the black mangroves helps to further consolidate the swamp sediments, eventually rendering them suitable for colonisation by the buttonwood, *Conocarpus erectus*, and other land plants.

In this way, three zones of mangrove vegetation are produced. The red mangroves with their distinctive prop-roots are submerged in salt water at high tide. Inland from these are the black mangroves, occupying the upper tidal zone in a swampy environment, and then come the buttonwoods, which are typical of the landward fringes of swamps. In some islands a zone of the white mangrove, *Laguncularia racemosa*, occurs between the black and button mangroves. This is

Flowers of the red mangrove.

The pneumatophores, seen here poking above the mud, are aerial roots of the mangrove which are capable of absorbing oxygen from the air.

especially true where dry or sandy soils occur, but *Laguncularia* is found more commonly as scattered individuals among the zones of red and black species.

Although mangroves seem to have an 'easy' life this is not really the case. For instance, although they grow in waterlogged or damp conditions the salt in the water makes it difficult for the mangroves to use it. Thus they are in danger of losing too much water by evaporation (transpiration as it is called in plants) from the leaves. As a result they exhibit a number of features which are more usually found in plants from dry conditions, in particular thick leaves with heavy waxy cuticles. It is this cuticle which gives the leaves their characteristic sheen. It is also significant that mangroves do well in conditions of high humidity and cloudiness, factors which reduce transpiration. Interestingly, these same factors seem to be of importance in the development of tropical rain forests. The waterlogged soils also make it difficult for the roots to 'breathe' and as a result the black mangroves, as has been mentioned earlier, possess pneumatophores which grow up into the air. It is a characteristic of demanding or difficult habitats that the range of successful species is restricted and again this is borne out by the handful of tree species found in these swamps. However, although it may be a difficult habitat for the trees it certainly is not so for the animals that thrive there.

There are three zones of mangrove vegetation.

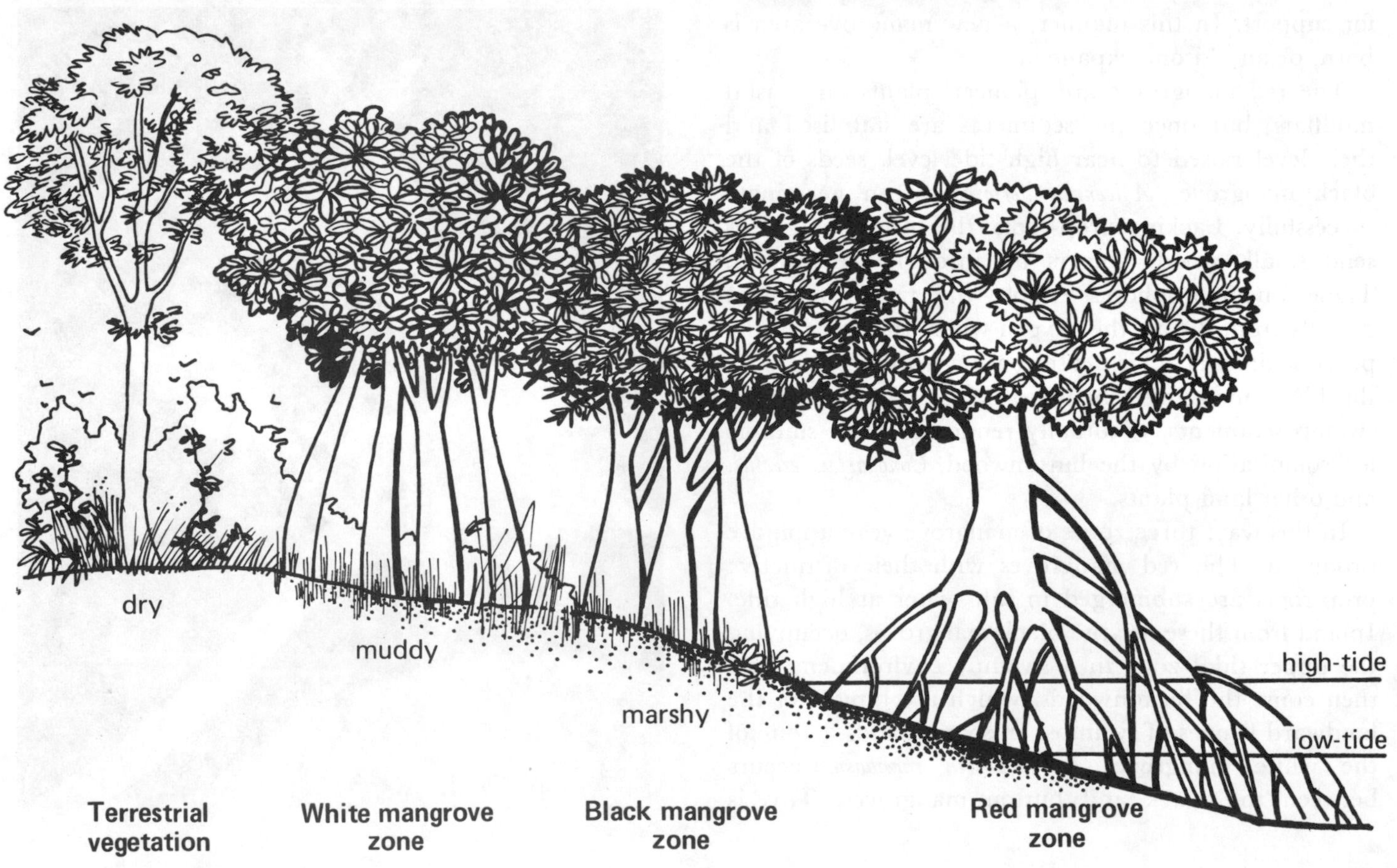

The Mangrove Community

On close study a variety of living things is found in the mangrove areas. This community of interdependent organisms makes mangrove areas among the most productive in tropical regions.

Nesting and roosting in the upper limbs of the mangrove trees are tropical birds such as the cattle egret and little blue heron. Nearby, spiders spin webs among the leaves to catch the sandflies and mosquitoes that breed in pools of dark water below. Other insects such as wasps and dragonflies dart busily above the damp roots. Ever watchful for a meal of insects are the anole, iguana and gecko lizards, slithering about the branches. Tree snails, *Littorina*, crawl along the tree trunks, leaving behind their silver trails. Certain species of crabs, rarely found in any other environment, inhabit this tangled world of roots and branches. They are predatory crabs like *Goniopsis*, *Pachygrapsus* and *Sesarma*, and the tree crab *Aratus*, which feeds on leaves, bark and algae.

At the water line clusters of oysters cling to the Red mangrove roots. There are two common species, the mangrove oyster, *Crassostrea*, most often found in brackish water regions, and the flat tree oyster, *Isognomon*, of more saline habitats. Attached to the roots, their shells are parted so they can draw in the nutrient-rich water which they strain for food particles.

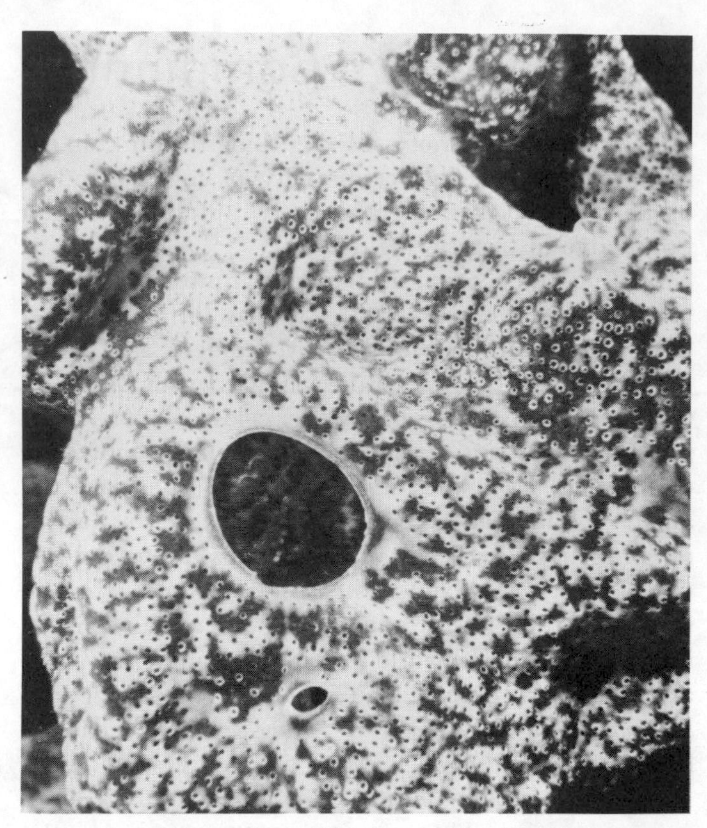

Colonies of sea squirts, sponges and sea anemones cling to the submerged mangrove roots.

If properly prepared for the table, oysters of both types are very tasty, so these shellfish are much sought after for food. Fishes with strong teeth, like puffers, also find them a perfect meal and *Murex* snails drill through the shells to get at the meat. Starfish are capable of prying the shells apart and inserting their stomach to digest the oyster flesh. Some birds feast also on the mangrove oysters, by poking their sharp bills in-between the shells and severing the muscle that holds them closed.

By far the most colourful growth on the roots is that of the sponges. Orange, red, blue and yellow encrusting species compete for space here. These animals are colonies of single cells which draw water into microscopic pores, filter out the food particles, and eject the water through larger holes which are visible to the naked eye. The sponges grow largely below the water line of low tide, where many other types of marine animal occur. Colonies of transparent tunicates cling to the roots in this zone. Tunicates, whose popular name is 'sea squirts', are mostly small animals looking like elongated bubbles. They siphon water through the body, feeding on the microscopic food particles it contains. Although these creatures are found on the open reefs as well, they exist in greatest abundance in the mangrove.

Small sea anemones attach themselves to the roots, their long tentacles spreading to capture food organisms which are then transferred to a central mouth. Larger anemones may be seen attached to the sides of lagoons just below the tangled root network; the large solitary species *Condylactis gigantia*, found here frequently, is also a common reef dweller.

Back among the roots, well hidden from view, juvenile fish of many species live until they are mature enough to migrate to the open sea or to neighbouring reefs. Young butterflyfish, angelfish, tarpon, snappers and barracudas peer out from the tangled roots, a perfect refuge. Sharing this nursery are juvenile spiny lobsters and shrimps, also destined to take up life in the open sea when they are older.

Resting in and on the mud below the mangroves are worms, small crustaceans and fat sea cucumbers. Perhaps the most bizarre marine resident of mangroves is the large jellyfish *Cassiopea*, which can be seen resting on the bottom of lagoons and water channels. In the Caribbean region there are two species of these jellyfish; *Cassiopea frondosa* and *Cassiopea xamachana*. They lie upside-down on the muddy bottom, pulsating; their complex branched tentacles spread upward to capture food particles floating in the tea-coloured waters. These jellyfish have a life span of one and one-half years and the largest specimens reach a diameter of 30 centimetres. They are normally absent from estuarine mangrove areas where salinities are probably too low.

Along the side of inlets and channels, short-spined sea urchins of the species *Lytechinus variegatus* cling to roots and fallen leaves, feeding on the abundant algae. Sea urchins are grazers, both here and on the open reefs, being especially common in beds of the turtle grass *Thalassia*, which may grow among the mangrove roots. A delicate and beautiful creature is the shell-less snail called a nudibranch or sea slug. Nudibranchs are seen infrequently on coral reefs and rocky shores, but are abundant among the roots in mangrove swamps. The most common species, *Elysia*, is coloured pale green and blends perfectly with its background of fallen leaves and marine grasses.

On the drier parts of the swamp away from the water's edge where the ground is covered with the black mangroves' pneumatophores there often live teeming hundreds of fiddler crabs (*Uca* spp.). These charming little crabs live in burrows often two or three centimetres in diameter. The name derives from the male of the species which has one claw much enlarged, and often brightly coloured. In display behaviour, this claw is waved across the crab's 'face' almost in the manner of a violin bow and it is this that

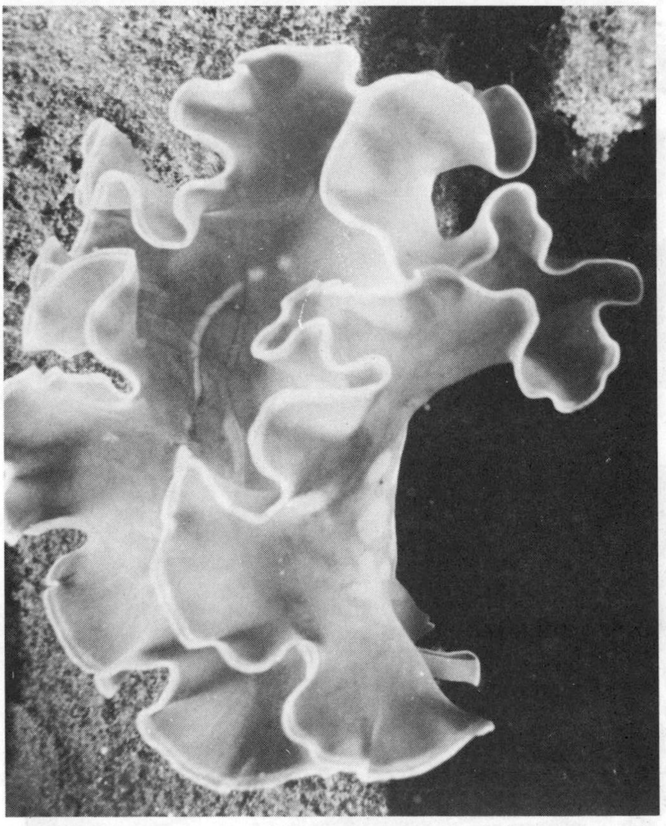

Sea urchins and sea slugs are common inhabitants of the swamp.

gives rise to the 'fiddler'. The display claw is so large it cannot be used for feeding and is simply a status symbol to frighten other males and to attract females (there is some doubt about its efficiency here!). Occasionally males will endulge in fights but more usually the matter is settled by a bout of claw waving. The females, with typical good sense, retain two small, but useful claws which are both used for feeding. The food is often tiny pieces of detritus picked from semi-solid mud surfaces uncovered when the tide is out.

Leaf fragments, or 'detritus' are fed upon by small creatures, which are in turn eaten by larger ones. Involved in this feeding pattern are shrimps, crabs, small fish and other marine life.

Decaying leaves form a nutrient rich banquet for tiny swamp inhabitants.

The Mangrove Food Web

The productive nature of a swamp is based on a pattern of inter-dependence among its great variety of life forms. This feeding pattern, based on who eats who, is a balanced system that sustains life here in the mangrove and affects the offshore reefs and other habitats, and is spoken of as a 'food web'.

This intricate food web starts with the mangrove leaves. Their lives spent, they fall to the mud below or into the water where they become coated with bacteria and fungi. Gradually over the months the leaves are broken down by these decomposer organisms into smaller and smaller pieces. These fragments of organic matter are now referred to as detritus.

Attracted to this rich banquet are shellfish, shrimps, crabs, worms, insect larvae and fish. In some mangrove areas, leaf detritus accounts for up to 90 per cent. of the

Maturing fish, lobsters and other animals gradually migrate from their mangrove 'nursery' to the open reef zones.

Larvae (newly hatched fish and shellfish) migrate from open reef areas to the still, protected waters of the mangrove to spend the early part of their lives.

diet of such animals. These detritus feeders in turn become the prey of several dozen species of juvenile fish and of larger invertebrates. Much of the particulate matter produced from the breakdown of mangrove leaves, fruit, flowers and twigs is transported out of the mangrove areas and forms the base of other food chains in other habitats. Those who view mangrove areas as wasteland to be filled with dredged material for conversion to building lots or agricultural land are blind to the role that mangroves play in the overall balance of nature and the productivity of island ecosystems.

Along the untouched mangrove coast man can study many varieties of life.

The Importance of the Mangrove

As hinted in the previous paragraphs, mangrove areas are important to man in several ways. It has already been explained that mangroves are land builders. As the colony of red and black mangroves increases, broken limbs, leaves and sediment accumulate and decay of roots in the soil leads to the formation of peat. Shells of shellfish, crabs and other animals are deposited also, and in some areas coral rubble is trapped among the prop-roots, building the shore-line up to a level where coconut palms, sea grape or other littoral plants may take root. In this way the land is extended slowly to seaward.

Mangroves grow most commonly on sheltered shores but in stormy weather the trees along the coastline become a vital buffer zone, protecting the land behind from erosion by high waves and hurricane flooding. It is likely also that mangroves filter off some of the land derived pollutants, thus protecting coral reefs and other littoral habitats from damage.

Perhaps the most important role of the mangroves is the production of a rich variety of food organisms. Many of these, such as fish, oysters, mussels, conch, crabs and birds can be harvested in the swamps and the production of protein from these sources is higher than from equivalent areas of agricultural land. The swamps provide food and shelter for a variety of creatures, including juvenile fish and crustaceans, destined to populate coastal and offshore areas where they will be harvested eventually during fishing operations.

It is obvious that we must learn to place a much higher value on this mangrove ecosystem which plays such an important part in the natural balance of life on and around the Caribbean islands.

Sandy Shores

The Caribbean is rightly famous amongst tourists for its beaches, be they golden, white or black. It is on these beaches that the visitor spends much of his time. These are the beaches upon which the fisherman pulls up his boat and upon which children play. However, from the point of view of their plant and animal life they seem at first to be dead, or almost so. As we shall see this is not really the case.

Sand is composed of small pieces of stone or shell and its colour depends on its origin. The pure white sands are derived from crushed coral; an admixture of shell fragments will colour the white sands yellowish or even brown. The famous black sand beaches of some of the eastern Caribbean islands are composed of tiny fragments of black volcanic rock.

Sandy beaches are very varied depending mainly on the amount of wave action. Sheltered bays tend to have gently sloping beaches of fine sand, grading into mud, silt and mangrove swamp. Exposed beaches on the other hand are often steep and composed of coarse sand with perhaps stones or boulders as well, and often grade into rocky or stony beaches. There may or may not be an offshore coral reef. Where this is present, the beach is mostly protected from large waves, as these break on the reef. This is particularly important from the conservationist point of view as clearly anything which tends to kill or destroy the offshore reef will often lead to drastic changes in the beach profile or perhaps to its complete disappearance. Too often ill-advised contractors have dynamited reefs without realising the consequences of their action.

Life in the Sand

Life in sand presents problems but also confers advantages on its inhabitants. The problems relate mainly to the shifting nature of the substrate.

Plants cannot attach themselves and large seaweeds are absent. Animals also cannot become fixed in the way that limpets and barnacles do on rocky shores or reefs. It is difficult or impossible for many animals to find homes; for example, crevice-living creatures so common under stones are not found in sand. Even the maintenance of a burrow may be difficult. The nature of the terrain also means that any creature who ventures forth on the surface of the sand is clearly visible to predators; gone are the hiding places of the mangroves and the reefs. The major advantage that sand dwellers gain is, paradoxically, protection. If they can manage to survive in, rather than on, this difficult substrate then they are invisible and often beyond the reach of many predators.

However, life in sandy beaches is never very rich, partly because of the difficulties and partly because the material to support the community must come in from outside. There are a few microscopic plants living on the individual sand grains, or even on the surface of the sea bed in quiet places, but much of what is consumed has to be washed in from outside. Most of this material will be either plankton or detritus so that the inhabitants of sandy beaches are mostly either those which sieve out these small food particles or those that prey upon the filter-feeders. Some gain their nutrients by eating the sand and digesting what little organic matter it contains, as earthworms do with soil.

Often the only indication of life below the surface is the presence of the open mouths of burrows. Some animals such as crabs may use these burrows as homes, and venture out in search of food. For others, who may live permanently in the sand, burrows are a connection with the world above the surface through which sea water is often drawn, bringing oxygenated water and food to the inhabitant. This is the case for many bivalve molluscs. Some worms may go on further than a burrow and build themselves a tube of sand grains cemented together with mucus into which they can retreat.

The relatively featureless sandy sea bed hides many burrowing and digging forms of animal life. Often all that is seen are their burrows.

The two-spined starfish, *Astropecten duplicatus*, lives in sand. It can bury itself quickly using its spines and pointed tube-feet.

Occasionally the large starfish *Oreaster reticulatus* will be seen moving over the sandy surface.

Filter and Suspension Feeders

Animals which feed on small food particles are usually divided into filter feeders and suspension feeders. The former usually create a current of water which they then sieve, collecting edible material of a particular size, other larger or smaller particles being rejected. Suspension feeders are often much more passive, simply holding out some large collecting organ and catching whatever the current brings them.

Many bivalve molluscs are filter feeders. They include *Donax*, sometimes called the chip-chip, which is very common on some exposed, surf-beaten beaches. This animal can be collected by hand and eaten either steamed or in soup. The shell is easy to recognise as its edge is milled like a silver coin and the small serrations can be easily felt with the finger nail. Other common sand dwelling bivalves include the pen shell, *Pinna*, which may reach twenty centimetres in length, various lucines and the beautiful tellins.

Various worms living in sand are also filter feeders. Especially noteworthy, are the beautiful tube-building fan worms. They have a colourful circle of tentacles which they hold out like an upturned umbrella to catch food particles. This delicate net is rapidly withdrawn into the tube at the slightest disturbance.

Pinna, a large bivalve, lives with its pointed end deep in sand while the open upper end of the shell just protrudes above the surface.

The large heart urchin, *Meoma ventricosa*, lives in sandy substrates at about 14 metres. Alive it is blackish and covered with short spines but its shell, when cleaned, is pure white and shows the typical echinoderm feature of five rays.

Flat sand dollars, *Mellita sexiesperforata*, live just under the surface. If removed and placed on the surface they quickly dig themselves in again.

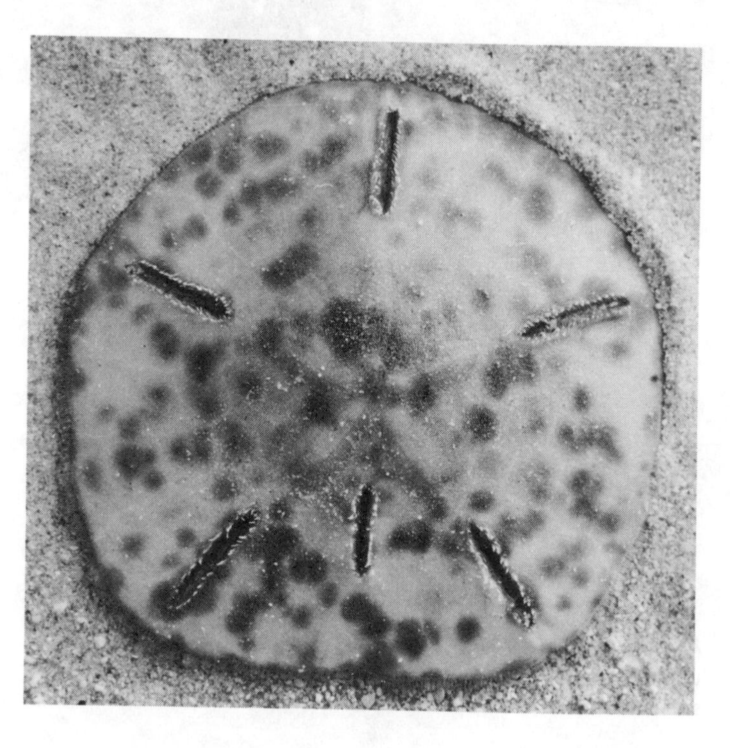

Sea urchins are quite common on sandy sea beds. They include the large dark red-brown heart urchin, *Meoma* and the flattened sand dollar, *Mellita*. Sand dwelling urchins feed in various ways, many of which are not well understood. In some cases they are suspension feeders, gathering up food that settles on or near them; some apparently search for food hidden in the sand while others may eat sand grains, digesting the organic matter from their surfaces.

Meoma frequently has small crabs living in among the spines of the underside where they gain a considerable measure of protection from predators.

Predators

The beautiful mollusc *Natica* lives in the sand ploughing its way along until it encounters a bivalve when it will bore a hole through the shell and consume the unfortunate occupant. The boring of the hole takes some time and is effected by the *Natica*'s rasping mouth parts aided by an acid secretion which softens the shell. It is not uncommon to find such bored shells washed up on the beach. The hole is easily identified as it gets narrower towards the inside of the shell and is usually not more than a quarter of an inch in diameter.

Another sand-dwelling predator is the box crab *Calappa*. This crab spends most of its time hidden under the surface of the sand holding its bizarre shaped claws neatly folded against its face. These strong claws are used to chip open the mollusc shells on which it feeds. They are also used in defence and can deliver a very painful bite (this information has been verified by the author!).

Sand Dwellers Above Sea Level

Many marine animals have evolved over millions of years to the extent that they can make use of the boundary between sea and land. This applies to all sea shore habitats. On most Caribbean sand beaches one will observe above high water mark the burrows of crabs. These burrows may be right on the beach, a little way inland or even far from the beach. There are three main types of crabs responsible for making these burrows. Those nearest the water edge are probably the easiest to observe. Most sunbathers if they lie quietly will observe the appearance from such a hole

Collecting animal and plant remains on the tideline can be very rewarding. This picture shows crab skeletons, limpet, snail and bivalve shells, sea urchins, a soft coral and sargassum weed all collected in about five minutes.

of a pale-coloured active crab which runs about the sand in an effortless way almost as if blown by the wind. These ghost crabs, *Ocypode*, are tideline scavengers, difficult to catch and capable of delivering a painful pinch if cornered.

Further inland are found two common species of land crab. These are the red coloured *Ucarcinus*, and the larger grey coloured *Cardisoma*. Both are edible and as they are largely nocturnal are hunted by torch light. In Antigua, ingenious wooden live traps are set for them. *Ocypode* is often 'torched' for bait by fishermen.

Tide Line

Beach combing is most rewarding on sandy beaches. Here at the high tide mark and at the water's edge one finds bits and pieces of animals and plants from deeper regions of the sea. Shells, of course, make up the most beautiful part of any flotsam collection. If there are coral reefs offshore or a rich sandy bay there is a real chance of finding decent specimens of the smaller conchs, star shells, small tritons, coweries, top shells and so on. Occasionally there will be a rarer musical volute or helmet shell. Sea urchin shells will also be common, often bleached white, while crab claws and lobster legs are attractively coloured in reds, blues and yellows. One great advantage of collecting in this way is that one is not killing the animal to take its shell. While any conservationist rightly deplores the taking of live molluscs for their shells no one objects to the beach comber collecting the many beautiful but empty shells thrown up on the sandy beaches.

Plant remains, too, can be fascinating. Many seeds and fruits of land plants float in the sea and so are washed from island to island; even from continent to continent. The best known example is the coconut and at the back of undisturbed beaches one frequently finds germinating fruits. Other seeds which are not un-

Fragments of dead coral cover a narrow beach.

Coconuts float in the sea and when washed up on a suitable beach will, like the one shown, germinate and grow.

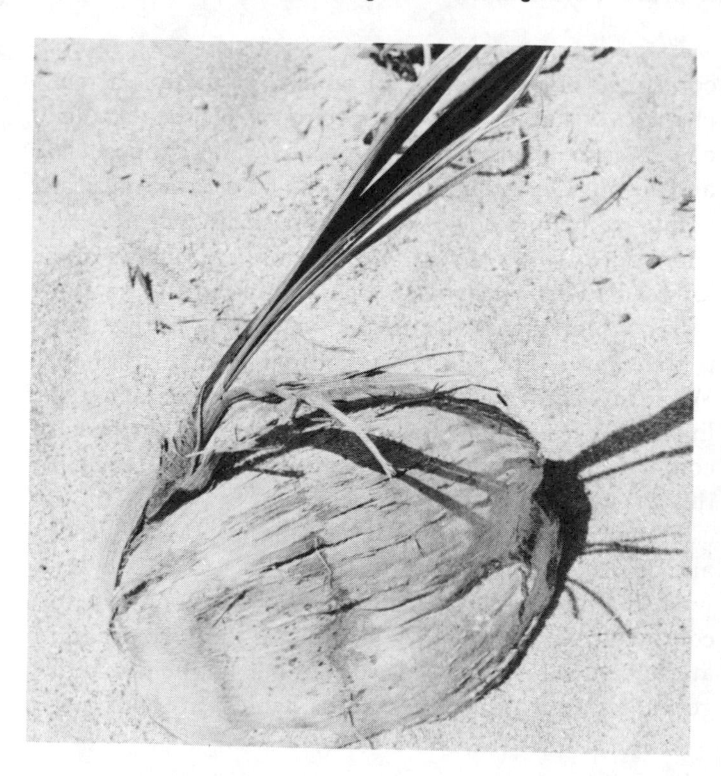

common are the shiny, grey, stone-like horse-nickers and various large beans. Seaweeds also are cast up and those which are well worth collecting are the various calcarious algae. These plants lay down a skeleton composed of calcium carbonate and long after they are dead and gone the shell remains, bleached white. One of the commonest genera is *Halimeda*.

Sea-grass Beds

This sandy bay in Antigua contains large sea-grass beds which show as the darker areas.

Thalassia testudinum is the most important tropical species of sea-grass.

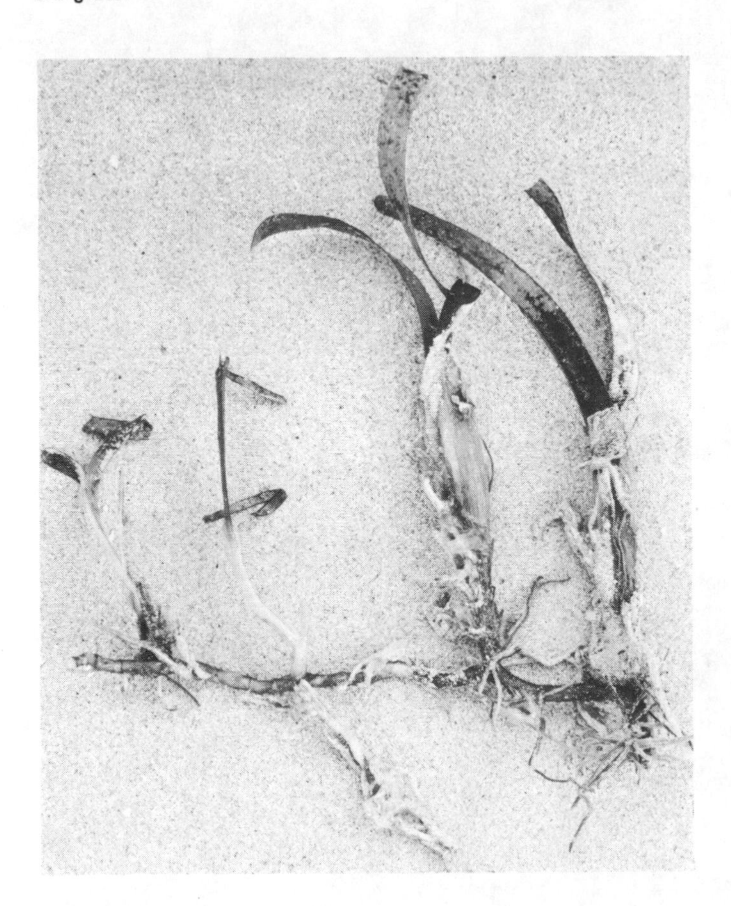

On shallow inshore waters one often finds beds of grass-like plants growing over large areas. These plants are unusual marine plants for, unlike the true seaweeds which are algae, the sea grasses are flowering plants. Two main genera of turtle grass, *Thalassia* and *Diplanthera*, are recognised. The latter usually grows closer inshore and can be distinguished in that beds of it look dark compared with the lighter coloured *Thalassia*. *Diplanthera* is also somewhat more tolerant of fresh water running off or through the beach. Neither can tolerate being uncovered for any length of time by the retreating tide. It is worthwhile making the point here that in the tropics generally there is not a rich inter-tidal flora and fauna and those more used to temperate regions will miss the jumbled masses of brown and red sea weeds that low tides usually reveal. There are two main reasons for this. Firstly, the relatively small tides mean that the total area covered and uncovered by each tide is small. Secondly, because of the tropical sun, the marine plants and animals of the inter-tidal zone are subjected to a tremendous drying and heating influence when uncovered; few are able to survive. The low nutrient level and active herbivorous habit of many animals reduces seaweed growth even further.

Sea-grasses not only have the above-substrate blades but also grow extensive underground creeping

Sipunculid worms are common inhabitants of *Thalassia* beds. This is *Phascolosoma antillarum*.

Sea cucumbers, snails and the common bivalve ***Brachidontes citrinus*** all burrow in the sand of sea-grass beds.

stems or rhizomes equipped with roots. As these plants only infrequently flower it is likely that dispersal is vegetative; that is by the breaking off and subsequent growth of fragments of the whole plant rather than by seeds. Indeed, in *Thalassia*, not only are flowers uncommon but all the male flowers appear at one time and all the female ones at another.

Because of their rhizotomous mode of growth seagrasses bind the substrate together and give it a solidarity which is lacking in sand alone. They grow best on sand mixed with coral fragments or stones and do not thrive where the particles are all of one size. This, together with the fact that the leaves are a rich source of food, results in grass beds being habitats with a rich and varied fauna. One study of a turtle grass bed showed that it contained 133 species of animals (excluding fishes) sixty of them found only in turtle grass. The nearby bare sand contained only 33 species of which only eight were found only in the sand. *Diplanthera* beds are also very rich in animal species but contain relatively few animals that are not found elsewhere. Many of the turtle grass animals live under the surface in burrows or buried in sand and they are best investigated by digging up some of the substrate and washing it through a coarse sieve (two or three millimetre holes). (Remember, however, that this destroys the habitat and should only be done in the course of serious investigation). In this way, one will discover a wealth of small creatures, especially small worms, crustaceans and bivalve molluscs. Particularly common among the shellfish are the bivalves *Codakia*, the tiger lucine and *Chione* the cross-barred venus, and the gastropod *Prunum*, a marginella.

Also common are brittle stars. These are five armed creatures and related to the starfish but have much thinner more spiny arms than their cousins. They live in crannies and are suspension feeders straining out edible particles from the passing water.

Above the sea bed some slightly more spectacular animals are to be seen. The pride of the turtle grass as far as the fishes are concerned are the sea horses. These bizarre fishes are occasionally to be seen clinging to the turtle grass by their tails. They are weak swimmers.

Where there are nooks and crannies it is not uncommon to find octopuses. These strange creatures are related to snails and bivalves and feed mainly on crustaceans. They are capable of walking over the surface of the sea bed or of swimming by the rapid ejection of water from the mantle cavity, a sack-like space containing the gills. When severely frightened they will squirt out a black ink that acts like a smoke screen behind which they can make their escape.

Under stones in the sea-grass beds one often finds the pencil urchin *Eucidaris tribuloides*.

Very obvious in amongst the turtle grass upon which they feed are large white sea urchins. There are two common genera, *Tripneustes* (sea egg) and *Lytochinus*. The sea egg is prized in Barbados where its roes are eaten as a delicacy (though it is usually fished not in grass beds but on wave washed reefs). These urchins can be handled relatively easily unlike their relative *Diadema* whose black spines are needle sharp and can deliver a painful wound.

Sometimes quite large conchs are to be found in the turtle grass beds and smaller ones are usually common. Indeed, the grass beds are nursery grounds for many animals. Like the mangrove swamp, the juveniles of many reef species are to be found here. In both places food is relatively plentiful. Very large predators are to a great extent excluded by the shallowness of the water, while there is plenty of cover in which small animals can conceal themselves. This cover is not sufficient for large animals but during the night the beds are invaded by bigger creatures, especially if there is coral nearby in which these animals can hide during the day. Thus parrotfish sally forth to eat the turtle grass while grunts go in search of invertebrates and other small animals for food. This exploitation of the beds is borne out by the high density of fish on reefs close to grass beds.

The commonest coral in turtle grass beds is *Acropora palmata* the elkhorn coral. In these quiet conditions the growths are often large, spreading out from a central 'stalk' like a giant table. These stands are usually isolated and are a natural gathering point for many fish.

Sea-grass beds like mangrove swamps and coral reefs differ from the open ocean in being highly productive. The leaves and rhizomes of grasses are the starting point for a large and complex food web based not only on the plant material itself but also the bacteria and moulds that bring about their decay. In contrast to most true sea weeds which are clean and shiny the leaves of turtle grass are covered with a growth of minute plants and animals. A scraping from such a leaf, if examined, under the microscope, reveals a whole new world. Most true seaweeds and some sedentary animals (e.g. sea fans) secrete toxic substances that prevent the accumulation of such a covering of plants and animals. Some of these substances may be of pharmaceutical use as antibiotics.

Rocky Shores

Rocky shores abound in the Caribbean. Because of their variability and complexity they provide homes for a wide variety of animals and plants.

In many places in the Caribbean there is no beach and the meeting of land and sea is a rock face. Such shores, although sometimes in sheltered areas, are often pounded by ocean breakers; such is the case for the east Coasts of many of the Lesser Antilles. Below the surface the animals inhabiting the rocks are in many ways similar to those of coral reefs. Indeed, if the wave action is not too great corals may thrive on top of the rocky substrate.

However, at the surface, the smooth wave worn rocks allow the settlement of many attached animals. Some of these are fixed in one place for life; others cling on while the surf beats on their backs but can move and feed when calmer conditions permit.

Into the former group fall the barnacles. These creatures are of tremendous importance on rocky shores in temperate regions but on many West Indian shores they may be hard to find. It is difficult for many people to realise that these small shelled animals cemented to rocks are in fact relatives of the active crabs and lobsters. However, they are indeed crustaceans as has been clearly shown by investigations of their life cycle. The eggs which are shed into the sea develop into tiny swimming larvae which live for a while in the plankton. At this stage they have the characteristics, including jointed legs and so on, of crustaceans, but if they survive, they settle on a suitable surface and change into the adult form. They retain their jointed legs, however, which are used to capture small animals for food. Other attached species include sponges, hydroids, anemones, tunicates and tube worms, although these are mostly truly submarine animals.

The 'hangers-on' are mainly molluscs; snails, limpets and chitons. The snails are mostly small and inconspicuous including the winkles. These tiny creatures often live high above the high tide mark browsing on algae and lichen on the rock surfaces. During the day when most of us go to such places the winkles are inactive. At night, however, they are out and about. The reason for this nocturnal activity is mainly to avoid predators such as crabs and birds which may hunt by sight, and to keep water loss by evaporation to a minimum. This loss would, of course, be particularly great in the heat of the day; indeed it is remarkable that the animals can survive the very considerable heating they must undergo on their sun drenched rocks. Although small, these common shells have a beauty of their own. Particularly attractive is the black and white *Littorina ziczac*, the zebra winkle, (for once the latin name is the most attractive).

The larger snails include the colourful and variable nerites, of which the bloody toothed nerite is perhaps the most spectacular. The area close to the opening of the shell is stained blood red so that its name is almost horribly appropriate. Usually, this species is found alongside another which is almost indistinguishable until turned over when the lack of the red stain identifies it.

The limpets are found further down the shore and may be of two basic types. These gastropods have uncoiled conical shells which are either complete or have a hole at the apex. The latter are separate from common or garden limpets and given the name keyhole limpets. As with the snails, their shells, although often small, repay careful examination. These creatures, like most of the snails are browsers, rasping the microscopic plants off the rock surfaces with a file-like tongue, the radula.

The chitons are molluscs whose shell covering is composed of eight jointed sections allowing them the flexibility to mould themselves to the rock surface and so withstand the force of the waves. So tightly can they adhere to the rocks that, if attempts are made to remove them, they often break apart before they detach.

Lower down rocky shores the fauna may be more varied and more like that of the submarine world, containing crabs (especially grapsoids), urchins and sea

The short spined black urchin, *Echinometra viridis*, lives in pools and bores shallow holes in rocks. This gives it some protection from the waves.

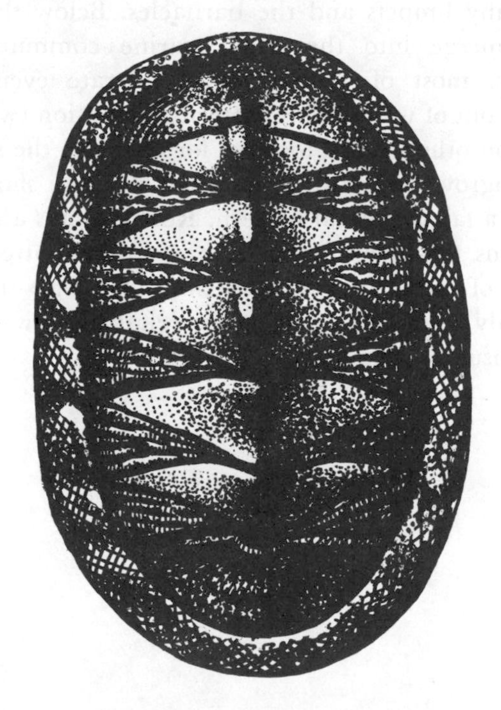

The chiton, *Chiton tuberculatus*, lives on rocks which, while often uncovered at low tide, are always covered at high tide.

Grapsis grapsis is a common and colourful crab which scurries about the rocks just above water level.

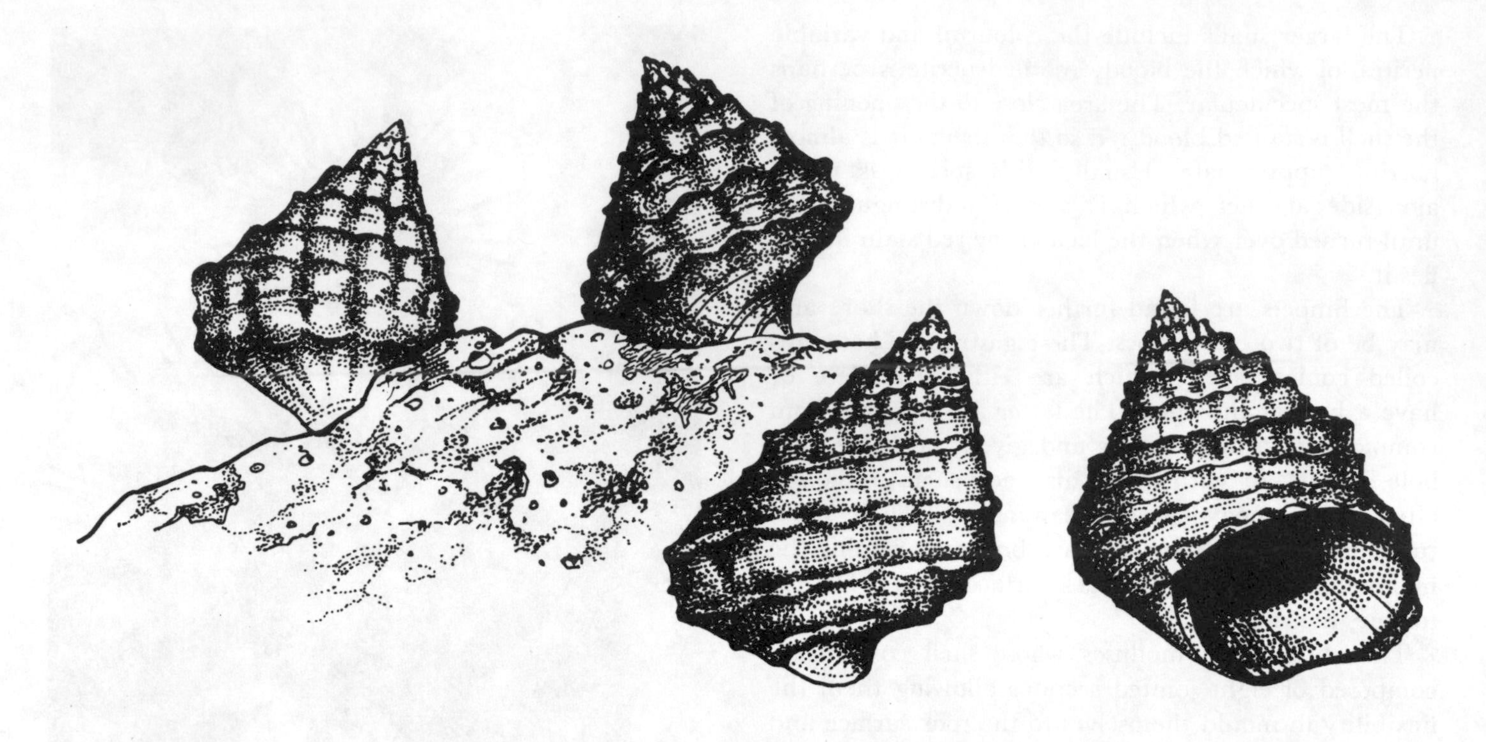

High above the sea, washed only by spray from breaking waves, are the tiny snails, *Nodolittorina tuberculata*, the common prickly-winkle.

cucumbers. One of the most attractive crabs is the calico crab, *Eriphia*. Although quite small, five centimetres or so across, he is a striking creature. The back is brownish-green but his claws are bluish above, yellow beneath and covered in chocolate brown warts. The tips of the pincers themselves are dark brown also. In the pools will be many small gobies and other fish. On surfaces constantly wetted seaweeds are often to be found in profusion, though never as great as in the temperate regions.

Rocky shores often show excellent examples of zonation, that is, the beach is divided horizontally into bands or zones each with its characteristic animals. Thus the highest zone, splashed only by spray, will contain such snails as the common prickly-winkle. Lower down where the rocks are regularly wetted by the high tides come the nerites and the littorinas. Below this comes a mid-tide zone which is usually submerged for half of the time or perhaps more. Here live many limpets and the barnacles. Below this, the zones merge into the truly marine community of animals, most of which cannot tolerate even brief periods out of water. This pattern of zonation (which is found on other shores too; see for example the section on mangrove swamps) varies from shore to shore but follows a fairly general pattern. Rocky shores are ideal situations for investigation into zonation patterns by groups of naturalists or students. The high shore is especially suitable as it contains relatively few species and is usually accessible.

Coral Reefs and their Inhabitants

Living polyps of the star coral, *Montastrea*, begin to emerge from their stony cups.

The tentacled polyps extended fully to capture food particles in the water.

A coral reef is one of the most complex communities of plants and animals in the world. The richness of the flora and fauna on the reef contrasts with the relatively barren seas surrounding it. However, we should not forget that the reef is a delicate thing whose life can be destroyed by quite small changes in the surroundings.

Pressures on the coral reefs of the Caribbean are growing year by year and there is an urgent need to conserve these valuable natural reserves. But for these reefs inshore fisheries would suffer and beaches would disappear under wave action.

Reefs are a tourist attraction and many visitors regard a glass-bottomed boat trip, a snorkel or a scuba dive session as the highlight of their Caribbean holiday. More and more the region will be forced back to its own resources and thus we cannot afford to damage or destroy reefs. Despite this, it is unfortunately true that destruction of coral reefs is taking place throughout the Caribbean. Sometime the destruction is in the name of progress as when deep water harbours are built or hotels constructed. Sometimes it is through ignorance as with the yachtsman or fisherman who smashes coral with his anchor. Sometimes, also, it is cynically and knowingly done by collectors or beach developers. In some places this process of destruction is being slowed or stopped by legislation and protection but the laws are often difficult to enforce. Moreover, in a region such as ours, even conservationists must realise the existence of competing claims; for instance, a deep water harbour may well benefit the local community a great deal. Perhaps, however, it could be constructed with the biological factors in mind. One hopes that in the future governments and those in control of development will seek out and incorporate the views of the conservationists into their policies.

On the constructive side there are a number of attempts now being made to create artificial reefs. This is being done by dropping wreckage and rubbish, old car bodies, tyres, even worthless hulks, into the sea at suitable depths. These objects are quickly grown over by reef organisms and become productive areas. Such projects are relatively cheap and have the dual advantage of reducing pollution on land while increasing the amenity and production potential of the sea.

The most important creatures on the reefs are the corals themselves. Corals are animals, albeit fairly simple ones. They are closely related to sea anemones and rather more distantly to jelly-fish. Coral organisms called polyps are rather like tiny sea anemones in appearance with a ring of tentacles surrounding a central mouth. Most coral polyps are withdrawn during the day but extended at night. However, this is not a hard and fast rule and coral polyps may extend in the daytime.

Corals can capture various tiny animals and plants that float in the sea. The animals, which are generally larger than the microscopic plants, are usually caught by the tentacles. These are armed with tiny sting cells called nematocysts which can be discharged explosively, throwing out threads which may both harpoon and poison the prey. (In large sea anemones, nematocysts may kill quite large fish while those of jelly fish and portuguese man-o'-war can cause very painful and sometimes dangerous stings to man.) The corals' prey is then transferred to the mouth. This may take place by a movement of the tentacles or by the prey being swept over the surface of the coral polyp trapped in a layer of mucus. This mucus covers the whole coral and is constantly being swept towards the polyp's mouth by the action of microscopic hair-like processes on the surface of the living tissue. It often acts as a trap for tiny floating plants which are passed in through the mouth and digested. The mucus also prevents larva of other animals settling on the coral and keeps it clean of sand and silt particles.

Large pieces of organic matter that fall onto a coral colony may be too big to be taken in through a polyp mouth. In this case digestive threads are passed out from the body cavity and surround the object, which is slowly broken down. Animals that indiscriminately catch any floating particles in this way are termed suspension feeders because they are feeding on suspended organic matter in the water.

Death of coral may occur in a variety of ways, but one of the commonest causes is over-growing. If two growing colonies meet the one tends to grow over the other. The covered coral will usually die under such conditions. Some slow growing species protect themselves from faster growing ones by thrusting out the digestive filaments mentioned earlier. These can kill the encroaching coral.

Some corals are composed of a single simple polyp while others are clearly colonies of many polyps. Yet other species are formed of polyps that are much deformed from the typical shape (e.g. brain corals.). In colonial forms new polyps simply bud off from existing ones. New colonies are produced by sexual reproduction. Eggs and sperm are released into the sea and the fertilized egg develops into a tiny swimming creature, the planula. Not infrequently the fertilization of the egg is internal. The sperms are formed from ridges inside the body cavity and shed out through the mouth. Some of these sperms are taken up through the mouth of polyps containing eggs. These are fertilised in the body cavity and some development takes place before the now fully motile planula is released. Experiments suggest that planulae at first swim upwards and towards the light. Later, however, they sink to the sea bed in search of a new home. The larval life usually lasts for only a few days after which the planula settles down and forms a tiny polyp. Many planulae are, of course, eaten as they swim in the plankton.

A remarkable thing about the coral polyp is the fact that it has living in its tissues thousands of single-celled algae (zooxanthellae). These do not harm the coral, indeed the relationship seems to be to their mutual advantage. A beneficial partnership such as this is termed a symbiosis.

The zooxanthellae obtain nutrients from the waste products of the coral polyps while the corals gain from the zooxanthellae the ability to form the hard calcium carbonate skeleton that is so characteristic of stony corals. We do not know exactly how this hard material is laid down but we do know that active photosynthesis by the zooxanthellae is essential for its formation.

A close-up photo of a living brain coral, *Diploria strigosa*, shows the stony valleys in which the soft coral animals reside.

Corals placed in the dark or whose zooxanthellae are destroyed are incapable of forming the typical skeleton. Because corals can only form their skeletons with the help of their zooxanthellae it is obvious that they cannot form skeletons in dark places. Thus corals do not thrive, as do some sponges and tunicates, on the undersides of stones. Nor do they grow well where suspended silt in the water cuts out the light. Sea water itself absorbs some light and as one goes deeper and deeper so it becomes more and more gloomy however bright the sun and clear the water. This means that corals will not grow well below a depth of about 60 metres and even below 30 metres the forms are relatively delicate. Massive corals are confined to shallow depths. The white sunlight is, of course, made up of light of various colours. These colours do not penetrate sea water equally. Red light especially lacks penetration, and at 20 metres red objects look black as there is no red light for them to reflect. This is the reason why many divers are astonished when they see colour pictures taken at depths with an electronic flash unit or flood light. For the first time perhaps they see the wealth of red tints and hues. The same effect can be obtained on a dive by taking a pressure-resistant flash-light which will supply the lost parts of the spectrum.

The process of coral growth is a relatively slow one. Although the slender branches of *Acropora cervicornis* may grow centimetres each year this is very rapid compared with other corals, especially very heavy solid forms such as the brain corals.

The other major contribution of the zooxanthellae to the coral is in the form of food. The tiny algal cells are able to capture the energy from sunlight and use it to make substances such as sugars. Some of these leak from the algae into the surrounding coral cells where, of course, they can be utilised as food. The amount of material passed to the coral in this way is not yet fully known.

As has been mentioned, the productivity of the reef is many times higher than that of the surrounding oceans and far higher than the nutrient levels of the seas would suggest is possible. This high productivity seems to be the result of at least two factors. Firstly, the protected zooxanthellae add considerably to the total photosynthetic capability of the reef; these algae may total three times the weight of the animal tissue of the coral. Secondly, the reef with its many different animals and plants tends to conserve its nutrients so that few are lost. Thus in the water close to the reef the nutrients, detritus and zooplankton levels may be tens or hundreds of times the levels in the open ocean, but these materials and organisms stay within the reef environment. Thus reefs seems to be a system which is very productive with effective turn-over and retention of nutrient resources.

We have seen that a coral polyp is like a tiny sea anemone which sits in a cup in its skeleton. This cup is not smooth inside, however, but has a knob or ridge rising from the centre of the bottom (the columella) while there are a number of flat vertical plates (the septa) coming from the outside wall towards the centre of the cup. Some corals are but a single polyp which may grow to be quite large. More usually, the polyp, as it grows, divides to form a colony. Sometimes this new polyp will form from tissue well outside the area of the mouth and tentacles. In that case the new polyps will be clearly separated, each having a separate cup. Sometimes, however, a new mouth may form within the area of the oral disc enclosed by the tentacles. In this case the new mouth may gradually grow away from the original and eventually form a new separate polyp or the separation may remain incomplete. *Eusmilia* is an example of the former type. Most specimens of this beautiful coral show polyps which are in various stages of this process. The latter case is shown in, for example, the brain corals. Here the oral discs surrounding the mouths never separate and the result is that the columella instead of forming a pillar takes the shape of a meandering ridge so characteristic of these corals.

All stony corals form their skeletons in more or less this way. Although there is incredible variation in skeleton form there is an underlying uniformity which can always be detected.

The gross shape of the corals also varies enormously. Perhaps the most beautiful are the delicate branching shapes of the stagshorn coral, *Acropora cervicornis*. To see this coral growing in beds like miniature forests with brilliant blue chromis swimming among them is perhaps one of the most beautiful sights awaiting the scuba diver. Impressive by their very bulk are the huge spherical masses of such corals as *Siderastrea*. Many of the brain corals can also reach massive proportions and here the convoluted surface patterns add further sculptural interest. Many smaller corals have a branched finger-like appearance. Two such corals, both common, are species of *Porites* and *Madracis*. In the beautiful *Eusmilia* the branches end in an almost flower-like single polyp. For stately beauty no coral can improve on the pillars of *Dendrogyra* which may reach up to two metres in height rising from the reef like massive stalagmites.

Staghorn corals, *Acropora cervicornias*, are abundant at all depths.

Pillar coral, *Dendrogyra cylindrus*, with a squirrel fish.

Some corals, especially those with an extended depth range, show variation of gross forms. Perhaps the best example is *Agarica agaricites*. This coral forms knobbly lumps in shallow water but at depths of 30 metres or so it takes the form of delicate flat plates.

Generally speaking corals are more delicate at depths below 25 metres. This is in part due to the low light intensity which reduces the photosynthesis of the zooxanthellae upon which skeleton formation is dependent. However, it may also be that heavier skeletal formations are an adaptation to the more turbulent water conditions nearer the surface.

The ability to form a hard skeleton, that will persist after the polyp dies, means that coral growth over the years can form solid rocky structures of very considerable size. The upper portions of these are, of course, the reefs themselves. Reefs may form in various ways and in various places. Different types of reef have often been given names such as fringing, patch, barrier, bank, etc. However, there is not always agreement about the use of these names and many reefs do not have a simple structure fitting one or other of the types.

Elkhorn corals, *Acropora palmata*, form the basic structure of most shallow reefs.

Barbados is largely composed of fossil coral rock. This cliff shows a large fossil coral head with flat fragments of *Acropora palmata* below.

Fringing reefs

These are the most usual type of reef in the West Indies. They are formed by coral establishing itself in the relatively shallow water at the edge of the shore and then gradually growing outwards into the sea. Such reefs come close to the surface of the water at their seaward edge and indeed waves may break on them. As the growth pushes out to sea the shore side of the reef tends to die and the dead coral compacts down to form a lagoon where depth may vary from a few to many metres. This lagoon may become filled with sand and support turtle grass beds (see p. 28). The front side of such a reef may fall steeply into deep water giving spectacular underwater scenery. The action of waves may erode gullies of considerable depth in the front face of such reefs. In other places similar gullies contain sand which gradually overflows from the lagoon.

Smaller gullies are often found in shallow water. They are cut in towards the shore and may have sandy bottoms. This type of underwater contouring is known as spur and groove, and is particularly attractive as the steep sides of the spurs form coral cliffs (although they may be only a metre or so high).

Each part of the reef has its own characteristic corals. For instance, the zone where the waves are breaking often is rich in elkhorn coral while the deeper front face of the reef has much of the more delicate stagshorn coral.

Barrier reef

This is a term which is sometimes used for fringing reefs which have formed parallel to a shore but with a fairly wide space between the shore and the shallow reef front. However, the term is best reserved for huge reefs found many kilometres off continental land masses e.g. the Great Barrier Reef off the east coast of Australia.

Patch reefs

These reefs may develop in water of medium depth on relatively flat sea bottoms if suitable conditions exist.

Reef structure is made more complicated by the fact that during the geological history the sea level in the Caribbean has not been constant and by the fact that many land masses have either risen or sunk. Thus in Barbados one can easily see a succession of reefs that have been lifted up one after the other out of the sea.

The building of coral reefs is not only brought about by corals themselves. Many seaweeds also form stony deposits and most importantly some corralline algae grow over coral debris cementing the loose pieces together to produce a solid firm substrate for further growth.

If one has the opportunity to examine a piece of coral rock, that is to say a fragment of a former reef perhaps hundreds of thousands of years old, one can see the many components. As well as the coral itself there is much compacted sand together with the skeletons of many marine animals and plants, especially molluscs, echinoderms and coralline algae. There may also be a sizeable contribution of foraminiferan shells. These tiny unicellular animals usually form a calcerious shell often like that of a minute mollusc. Although these seldom exceed a millimetre in size their very number may make then important rock formers.

Destructive forces also are at work in the reef. You will see in almost any piece of dead coral you collect on the beach a variety of holes and tunnels. These are produced by animals which bore into the coral to create homes for themselves. Investigations have shown that commonly 10 per cent. or more of the coral skeleton is destroyed in this way. Much coral is ground into sand and washed away to the sea bed or beaches.

The living coral, upon which the whole community depends, is a sensitive creature and changes in salinity, light and silt levels and some pollutants can result in the death of the reef. Although regeneration can occur it is slow indeed, as is the growth of the coral itself.

Many measurements of the ages of corals have been made and the smaller species seem to average seven years old (some species takes this long to reach sexual maturity). An Australian colony has been aged at 140 years but the larger West Indian species, especially those of the massive dome-like *Diploria* and *Montastrea* may be far older than this. Growth rates of just over one centimetre per year seem to be typical for these colonies and as many of them are over two metres in diameter this would make them over 200 years old.

Fire corals

There are in the reef some coral-like organisms that are in fact only rather distantly related to the true stony corals. The commonest of these are the fire corals. These creatures belong to a group called the Hydrocorallina. Their polyps protrude from the limey skeleton and can be seen with the naked eye as fine hair-like filaments. In fact, there are two distinct types

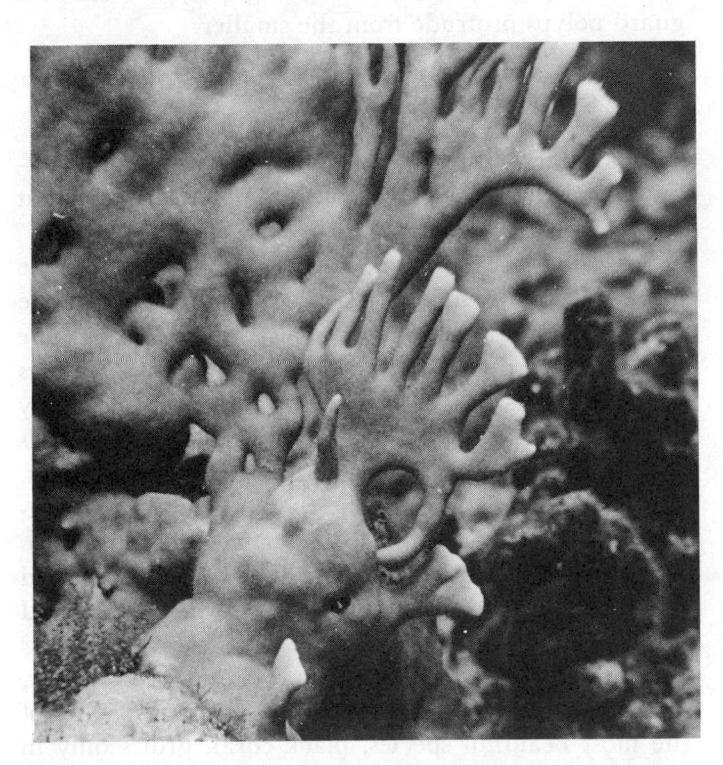

The fire coral *Millepora* sp. has a very variable and often beautiful form. However, as its name implies, it can deliver a painful sting.

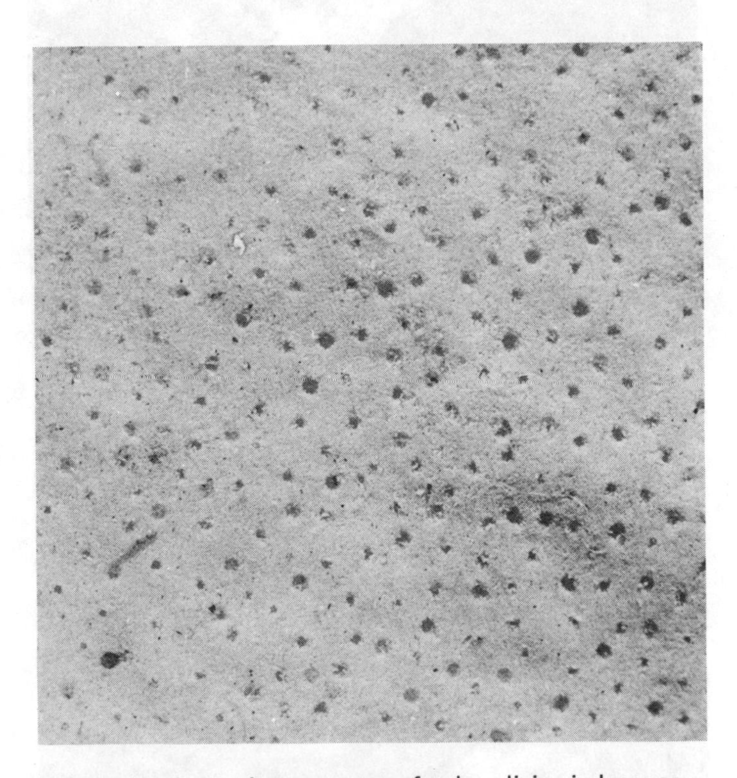

The fire coral contains two types of polyps living in large or small pores in the skeleton. These pores can be seen in this close-up photograph of a cleaned piece of coral.

A single sea whip frond extends its polyps; note that one polyp has closed around a captured food particle.

of polyps. Those that are visible appear to act as guards or touch sense organs. They have no mouth but carry short stinging tentacles along their length. The second type of polyp is much shorter and fatter. It has a mouth with four stinging tentacles around it. These are the polyps that capture and ingest the food, small planktonic creatures and bits of dead of animals. Inside the skeleton all the polyps are connected by a net of delicate tubes.

Fire coral is so called because of the burning sting it can deliver if touched. Many an unwary snorkeler has attempted to collect a piece with his bare hands, something he only tries once. It is the guard polyps which are mainly responsible for this sting.

The gross form of the fire corals is very varied. It tends to grow over objects; divers sometimes find discarded bottles that have been beautifully decorated in this way. In disturbed water the colonies may be rather stout, broad fans. In quieter waters the form may be much more delicate forming an elaborate filigree on the reef. On close examination the skeleton can easily be distinguished from the true corals by the lack of the typical rayed depression of the coral polyps. Instead the relatively smooth surface is peppered with minute holes of two sizes, the smaller barely visible to the naked eye. The larger house the feeding polyps while the guard polyps protrude from the smaller.

The soft corals

We have seen how the hard, stony corals are the reef-builders. However, the soft corals too play an important role in the reef community.

Sea whips, sea plumes, and sea fans comprise the major groups of soft corals. These all belong to the group Alcyonaria and have a structure that bears some similarities to their stony cousins. They are all colonies of small polyps borne on a skeleton, but this is usually horny with little or no calcium carbonate. Soft corals are pliant and can bend with the currents and waves. They feed at all times of the day and night, the tentackled polyps capturing food particles from the sea. Soft corals are fastened firmly to the bottom. They add grace and beauty to the undersea world, and should not be removed.

Both the soft and hard corals provide needed shelter, and food, for a great many marine creatures. One of the most beautiful species, black coral, grows only in deeper waters. It is frequently seen to provide a home for the colourful sponges that fasten to its base and countless brittle stars and other organisms. The

A red sponge and soft corals.

Banded butterflyfish, *Chaetodon striatus*.

A squirrelfish, *Holocentrus rufus*.

A coral reef comprising of sponges, sea feathers and seafans.

Trumpetfish, *Aulostomus maculatus*.

A fire coral growing alongside orange sponges.

A Serpulid, commonly known as the christmas tree worm.

Nassau grouper, *Epinephelus striatus*.

Queen angelfish, *Holacanthus ciliaris*.

Blackbar soldierfish, *Myripristis jacobus*.

Glasseye snapper, *Priacanthus cruentatus*.

Spotfin butterflyfish, *Chaetodon ocellatus*.

French angelfish, *Pomacanthus paru*.

A condylactis anemone and a diamond blenny.

Royal gramma, *Gramma loreto*.

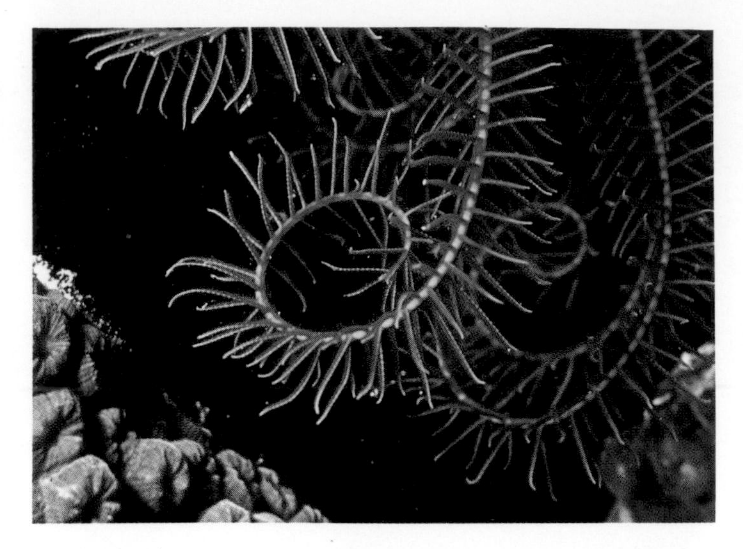

Crinoid.

A stoplight parrotfish, *Sparisoma viride*, taken at night.

Schoolmaster snapper, *Lutjanus apodus*.

collecting of black coral for the manufacture of jewellery has long been a controversial issue. Like any creature that has value for man, the black coral can be easily depleted if care is not taken to preserve it. It is worthwhile adding that traditional red or pink coral jewellery is made from an alcyonarian, *Corallium*. In this case the skeleton is mostly of coloured calcium carbonate. This animal is found most commonly in the Mediterranean Sea.

The Sponges

Other soft creatures that adorn the coral reefs include the sponges. The species found here are not the household variety, and are therefore of no commercial value.

Each sponge, no matter what its shape or size, is a living colony of single cells. The sponge 'feeds' by drawing water into its millions of microscopic chambers, straining out plankton, and ejecting the filtered water through the large holes that are seen easily with the naked eye.

Sponges are very primitive. They belong to the group known as Porifera, and basically they have evolved very little from their prehistoric ancestors. The largest species found in the Caribbean are the basket sponges, so big at times that two human beings could sit inside one. At the other extreme, the encrusting sponges are small colonies found covering the corals, or the roofs and walls of undersea caves. Sponges support themselves with thousands of tiny spicules made of protein, silica or calcium carbonate.

If you remove a living sponge from the sea its bad smell will soon cause you to wish you hadn't! So beautiful and colourful are the sponges that they should be left in their natural habitat, where they can be viewed by divers or riders in a glass-bottomed boat.

Sponges play an important part in the reef community. Like the corals, they act as a refuge for small fish, starfish, shrimp, and other tiny creatures.

The strange phenomena of the 'smoking sponges' can be viewed at times by divers. To reproduce, sponges emit clouds of sperm and eggs into the water, resembling smoke.

Some sponges are capable of giving a very painful sting to the unwary diver or snorkler. These include the aptly named do-not-touch-me and fire sponges. Most stinging sponges are red in colour and to be on the safe side it is best not to touch any red sponge as the toxin laden spicules of the stingers can spoil a day or

Sponges feed by drawing in water through microscopic holes. Nutrients are strained out, and the water is then ejected through the sponges largest openings.

A diver relaxes in a huge basket sponge.

two of your holiday or diving trip. It should also be noted by the diver that, because sea water absorbs colour, red usually appears either maroon or black at depths of more than five metres.

The Spiney-skinned Echinoderms

Echinoderms all have some type of internal skeleton and often show a five-rayed symmetry. There are five major groups, the primitive crinoids or feather stars, starfish, brittle stars, sea urchins and sea cucumbers.

The crinoids are relatively uncommon. The species most often seen is the black feather star sometimes found clinging to the reef or rock surface. Its ten feathery arms are adapted for sieving out small food particles from the passing sea water.

The starfish are more common on sandy sea beds than on the reef. Their arms are thick and fleshy and equipped with tiny sucker feet. These are used both in locomotion and often in holding their prey, small bivalves. Among the commonest species are pointed armed *Luidas* and the blunt armed comet stars *Linckia*. In some places the colourful and spectacular *Oreaster* is common but uncontrolled collecting has sadly depleted its numbers in many islands.

The brittle stars also have five arms but unlike the starfish these are thin, spiny and attached at the centre to a small fleshy disc. Most of the Caribbean species live under stones or in crevices in the reef but two species are not infrequently found in the open. *Ophiothrix* is usually encountered clinging to sponges or sea fans, using its sticky tube feet to catch passing food organisms. The bizarre basket star, *Astrophyton*, is also found clinging to sponges and sea whips. This animal has an extraordinary appearance as the five arms branch again and again to produce a tangled mass often quite brightly coloured and as much as one metre across.

The sea urchins are very common on most reefs. The most familiar is the long-spined black *Diadema*. This creature rasps the surface of dead coral and rocks, eating the algae and other small creatures. Its spines are many centimetres long and needle sharp being mainly composed of a single calcium carbonate crystal. The spines can penetrate deep into the flesh of an unwary foot or hand where they break off leaving a painful wound, made worse by the fact that the spines contain a mild toxin. The traditional West Indian treatment in such cases is the application of lime juice, but many hold that rum taken internally is almost as good! The forest of spines may sometimes serve as a refuge for small fish. In some islands, especially Barbados, the roes of the white sea egg, *Tripneustes*, are eaten as a delicacy. These animals thrive on surf swept reefs and in turtle grass beds. Other common urchins include the red-black *Echinometra* which bores depressions in rocks at the sea's edge and the beautiful pencil urchin with its thick stubby spines. Heart urchins, *Meoma*, are found buried in sand and although they are closely related to *Diadema* and the others, they have lost the latters' symmetrical spherical shape. They have a recognisable front and back, top and bottom. This is also true of the sand dollars, flat biscuit-like creatures which also live buried in the sand. The commonest species in the Caribbean is the six-hole sand dollar.

The sea cucumbers are also more asymmetrical than most of their echinoderm relatives. These animals have

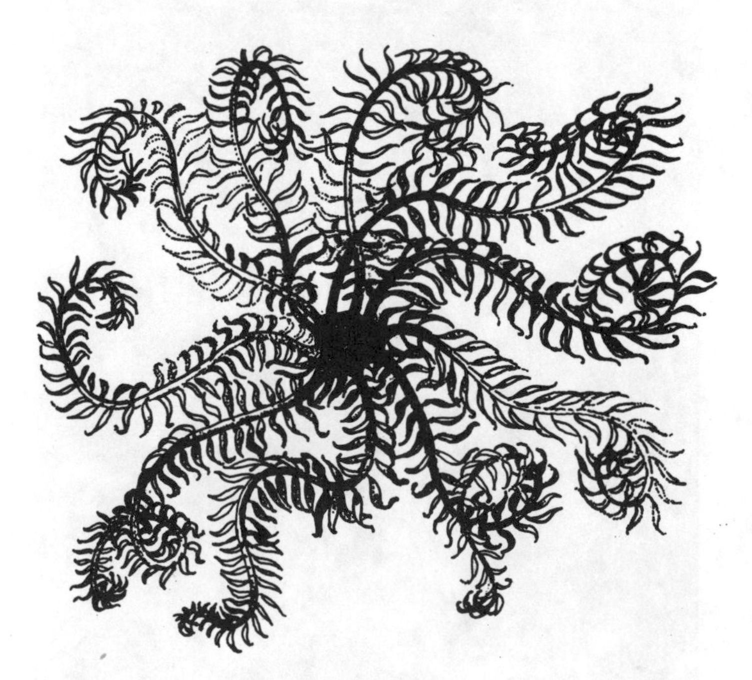

Feather starfish, or crinoid *Antedon*.

Sea cucumber, *Cucumaria*.

a much reduced skeleton and are soft to the touch. The reef species are usually small, perhaps ten centimetres or so long and live in nooks and crannies or under stones. On sandy bottoms *Actinopyga* is often quite common. This attains considerable size being up to 30 centimetres in length. Its upper surface is almost brown to black and has a warty appearance. The underside is smoother and lighter often blotched with brown and orange.

Worms of the Reef

The worms that have many bristles on their bodies are the polychaetes. They are all marine and, ironically, some of the most beautiful creatures on the reef.

Worms of the sea fall into four basic categories according to their living habits; free swimming, free crawling, tube dwelling and burrowing. This last category has many representatives on our sea bottom, but of course they are well hidden from our view.

The free-swimming worms are very small, spending their lives among the plankton.

Of the crawling variety perhaps the most significant on our reefs is the fire worm. It has rows of glass-like bristles down both sides of its body, a perfect defence mechanism. Anyone or anything touching them receives a painful sting.

The most spectacular worms on the reef, and the easiest to see, are the sabellids and serpulids. Often called feather dusters, the sabellids live in flexible tubes which they secrete. The smaller species often live clustered together giving the impression of a bouquet of flowers each about two to three centimetres across. The 'blossom' is actually a ring of tentacles used by the worm to trap floating food organisms. Larger species are more usually solitary and may be up to 10 centimetres across. These beautiful animals would appear to be rather vulnerable to having their tentacles eaten but touch one and you will see the lightening withdrawal of the worm and its tentacles into the tube. The tentacles not only respond to touch and vibrations but also are light sensitive; if a shadow passes over the crown it often produces the withdrawal response.

The serpulids have been named christmas tree worms. Each one lives in a secreted tube that penetrates the coral. This worm has a twin gill, two spiralled plumes that, like the feather duster's gills, act both as a breathing and feeding device, straining plankton from the water. At the slightest disturbance, these plumes disappear suddenly inside the tube.

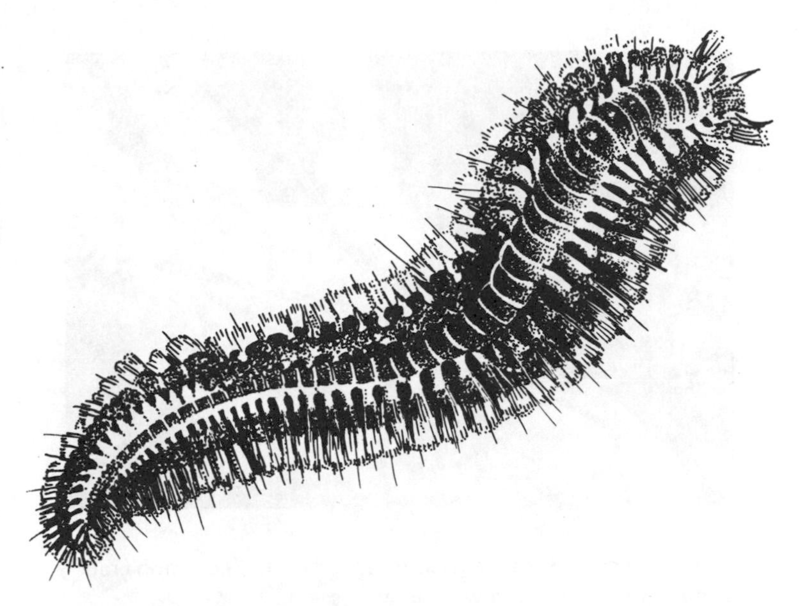

The fireworm, *Hermodice*, is well protected from predators.

The sabellid worms resemble feather dusters, hence their common name.

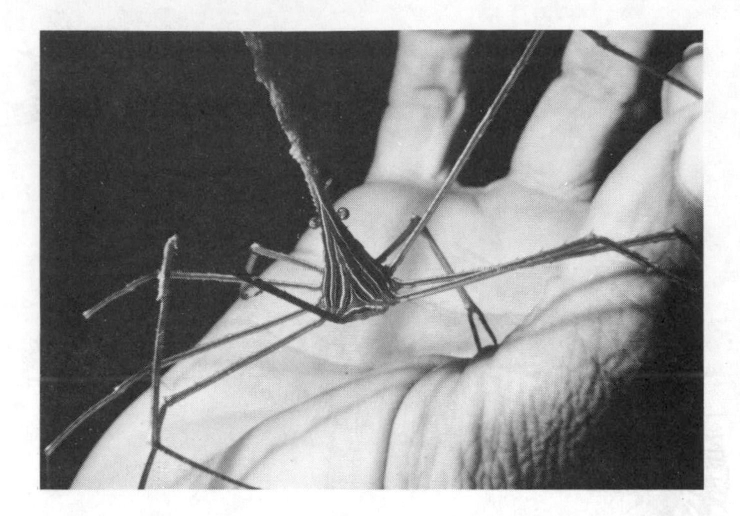

The arrow crab, *Stenophynchus seticornis*, is a common crab of coral reefs but not always easy to see.

This hermit crab is seen in his home, a West Indian top shell.

This hermit, *Petrochirus diogenes*, can grow to more than 30 cms in length. It often lives in old conch shells.

The flat-worms belong to a quite different and more primitive group than the polychaetes. One of the commonest species in the Caribbean is *Ppeudoceros paradalis*, a truly remarkable animal, basically black with bright yellow blotches to its centre and white dots around the margin. Divers and snorkelers are most likely to find it under stones.

The Joint-legged Animals

The marine creatures which have jointed legs are called crustaceans. They include reef dwellers like the crabs, lobsters and shrimps. They are characterised by hard, jointed outer skeletons, antennae, and claws. The most popular member of this group locally is the West Indian langouste. However, a large variety of crabs and shrimps inhabit the reefs as well.

Many species of crabs occur on or near the reef. The most spectacular is probably the brightly coloured arrow crab, *Stenorhynchus*. This tiny spindly crab has a white or brown back lined in darker colours, blue claws and legs spotted or banded in red. It is sometimes found with wisps of seaweed attached to its long pointed 'nose' possibly acting as either a food store or camouflage, or perhaps both. The sponge crabs, *Dromidia*, are more difficult to see for they have a piece of living sponge on their backs. This is usually hollowed out to fit the shape of the shell and held in position by an upturned pair of legs. If upon turning over a stone you see a piece of sponge apparently walking away you can be fairly sure it is one of these strange creatures. The swimming crabs are handsome and wide shelled. As their name implies, they swim strongly, using the flattened end of their last pair of legs as oars. These crabs are active hunters even catching live fish. They often show differences between the two claws, one being a light slender catching claw and the other a stronger crusher. The stone crabs, *Menippe*, are much more heavily built with massive black claws. In some places these crabs are eaten and often the fishermen break off the claws returning the crab to the sea where such clawless individuals often grow new claws. They have a roughened area inside their claw which when the crab rubs it against the shell produces a characteristic noise.

Hermit crabs are very common on the reef. These, unlike the true crabs, have a soft abdomen which is adapted for clinging inside discarded snail shells. If touched, the crabs will retreat into their home blocking the entrance with the larger of their claws.

Snapping shrimps are not uncommon under sea anemones. These little crustaceans have dissimilar but quite heavily built claws, and are able to produce a characteristic clicking noise with the larger one. More often found in sandy areas are the carnivorous mantis shrimps. They have curious claws which can be extended at great speed to catch their prey, often small fish or other crustaceans. The largest of these, *Lysiosquilla*, may be up to 30 centimetres in length.

While our reefs do not contain any species of shrimp large enough for the table, there exist a host of brightly coloured but very small shrimps, some of which have a most important role in the reef community. These are the 'cleaner shrimps' which remove parasites from the bodies of fish (see page 63). The largest member of this group is the banded coral shrimp with its red and white armbands. Its long white antennae, emerging from beneath a coral overhang, invite clients who need to be 'cleaned'. Some shrimp species live with anemones in the interesting relationship described on page 64.

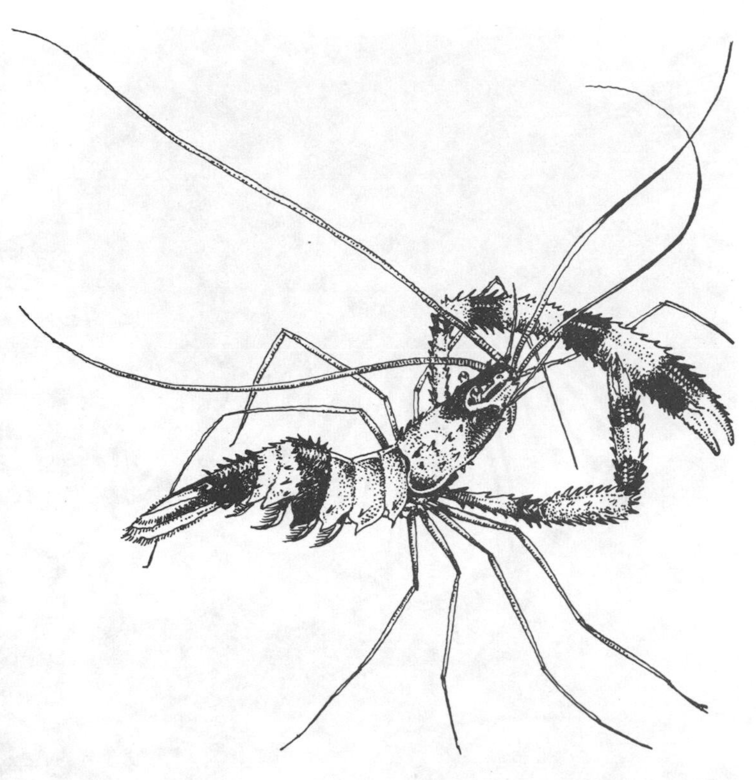

Stenopus hispidus, the banded coral shrimp, cleans fish of their parasites, advertising this service by waving its long, slender antennae.

The West Indian langouste or spiny lobster, *Panulirus argus*, was at one time plentiful in many parts of the Caribbean. However, over the years heavy fishing has depleted the stocks. It is a rare sight indeed to see the 60 centimetre, 15 kilogramme monsters that must have been relatively common. Usually today the lobsters caught average only about one or two kilogrammes.

The adult animal is a splendid sight, varying in colour from greenish-brown to dark red and having a spiny carapace and long antennae. The animal can walk over the sea bed using its legs or it can swim rapidly backwards by powerful curling movements of the abdomen which has a broad tail. It is the muscles of the abdomen that are the gourmets' delight. This animal lacks the massive claws of its cousins from northern waters but it does have large mandibles which are used for grinding up a variety of shelled animals. A Jamaican study showed that they mainly eat bivalve molluscs and also some crustaceans, plants and possibly fish. It is likely that they eat sea-urchins too, where there are available. They seem only to feed at night.

Lobsters breed all the year round in the Caribbean but mating is most frequent between February and August. They are able to breed when they are about 15 centimetres or longer overall and an average female may produce 850,000 eggs. When these have been fertilised they are held under the female's abdomen for about four weeks, at the end of which time they hatch to release tiny larval lobsters. These live first of

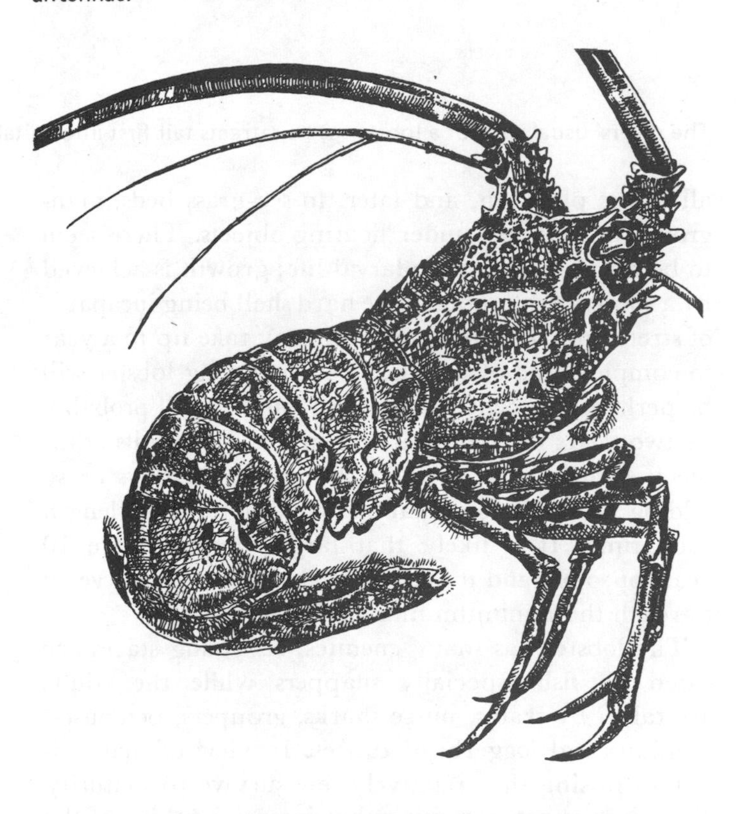

Although lobsters are delicious and therefore greatly in demand by both residents and visitors, their numbers have decreased as the demand has overcome the supply.

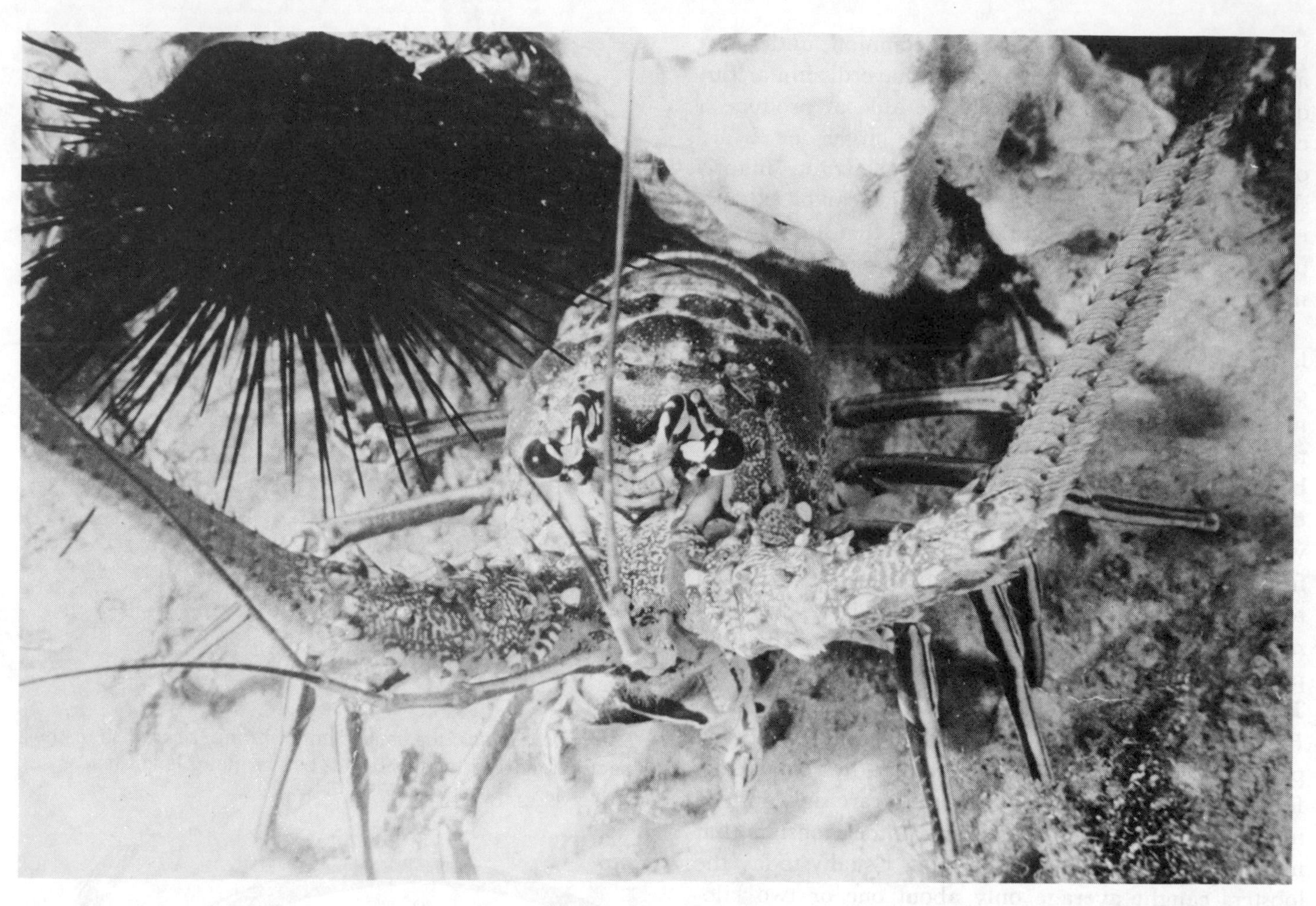

The divers' usual view of a lobster as he retreats tail first into suitable shelter.

all in the plankton, and later, in sea-grass beds, mangrove swamps and under floating objects. There seem to be many stages in the larval life; growth is achieved by a succession of moults, the hard shell being incapable of stretching. This pre-adult life may take up to a year to complete and at the end of it the young lobster will be perhaps five centimetres or so long. It will probably be two years old before it can breed. During its adult life it continues to shed its skin every 90 days or so adding probably less than one centimetre to its length each time. It is likely that large specimens are 10 years or so old and it takes probably three to four years to reach the minimum marketable size.

The lobster has many enemies. Its young stages are eaten by fish especially snappers while the adults are taken by skates, nurse sharks, groupers, octopuses, dolphins and loggerhead turtles. In view of that it is not surprising that relatively few survive to maturity. Equally it is not surprising that intensive fishing of the adults has resulted in some places in depletion of the stocks, reduced catches and small average sizes.

In many parts of the Caribbean legislation has been passed to control the minimum size of animal caught and there is often a closed season to allow breeding to take place (and also perhaps more importantly reduce the numbers of lobster caught). It is also often illegal to take females carrying eggs. Unfortunately, this legislation is often either inadequate or unenforceable and there is no doubt that some areas have been virtually 'fished out'.

Efforts are being made around the world to breed lobsters in captivity but although this has been achieved in one or two cases the cost at present makes it uneconomic even for this vastly expensive delicacy. The only answer to our problem today seems to be stricter laws, better enforced, combined with an effort to educate those responsible for the over exploitation. Perhaps one should end on a less pessimistic note and say that there are still one or two areas of the Caribbean where biologists are convinced that greater exploitation can take place without damage to, or depletion of, the stocks.

Shellfish

Molluscs are those marine creatures whose beautiful, intricate, and often colourful, shells are most often found on the beaches. The shelled molluscs are divided into two main groups, the first with a single shell which often shows coiling is the snails. The second is the bivalves which have two shells joined together by a ligament. The living or recently dead specimen has both shells but often they come apart after death and the beach comber will more frequently find only single valves. The molluscs also include the sea slugs, squids and octopuses which have either no shell or one which is much reduced and may be internal.

Usually one finds more molluscs on sand or eel grass than on the reef itself but we will deal with them altogether here for convenience.

Shells come in a great variety of shapes. Of the snails, the simplest are the almost conical limpets which cling to rocks around the low tide mark and below. Some are found to have a slit in the apex of the shell and are called keyhole limpets. Of a more typical snail-like appearance are the common West Indian top shells. These spirally wound shells are blotched black and white on the outside and are lined with iridescent mother-of-pearl. Large specimens may be ten centimetres across. Another very attractive top shell is the chocolate lined top shell, *Calliostoma javanicum*, which is very straight-sided and pointed with extremely regular whorls.

Some snails have a proboscis or siphon that they hold out in front of them as they move. This is essentially a sensory organ. The shells of snails with such a structure have a syphonal canal at the anterior end of the opening. Such snails include the murexes, conchs and tritons. The former have shells richly ornamented with spines and ribs while the latter two groups are among the most beautiful of West Indian shells.

In the coweries and their relatives, the aperture to the shell becomes large and slit-like as the shell grows and in the adult state it stretches the full length of the ventral side. In life the soft part of the animal frequently spreads out almost enclosing the shell. In the flamingo tongue, *Cyphoma*, the skin of this soft tissue is pigmented orange with numerous black rings. This striking animal can often be found feeding on soft corals.

Another attractive group of snails are the cones. These have a long aperture set in a conical shell. The coiled spine part is often low or almost flat. These

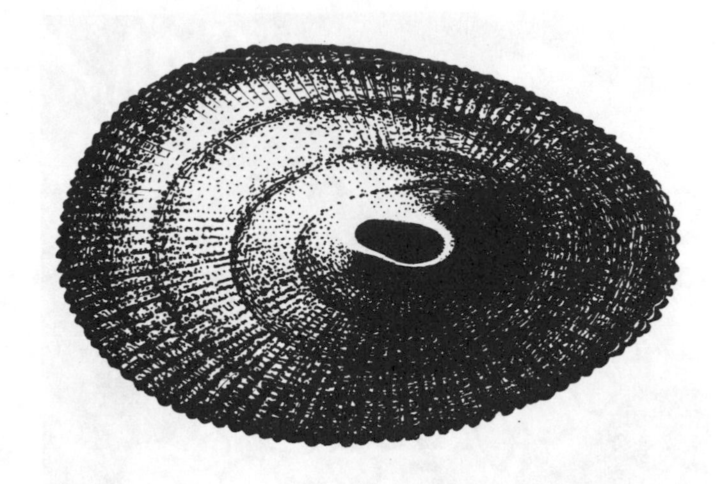

The limpet *Diodora*.

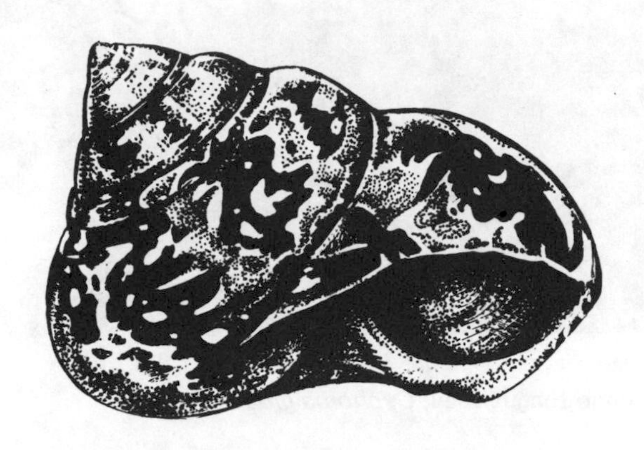

West Indian top shell, *Cittarium pica*.

Reticulated cowrie helmet, *Cypraecassis testiculus*.

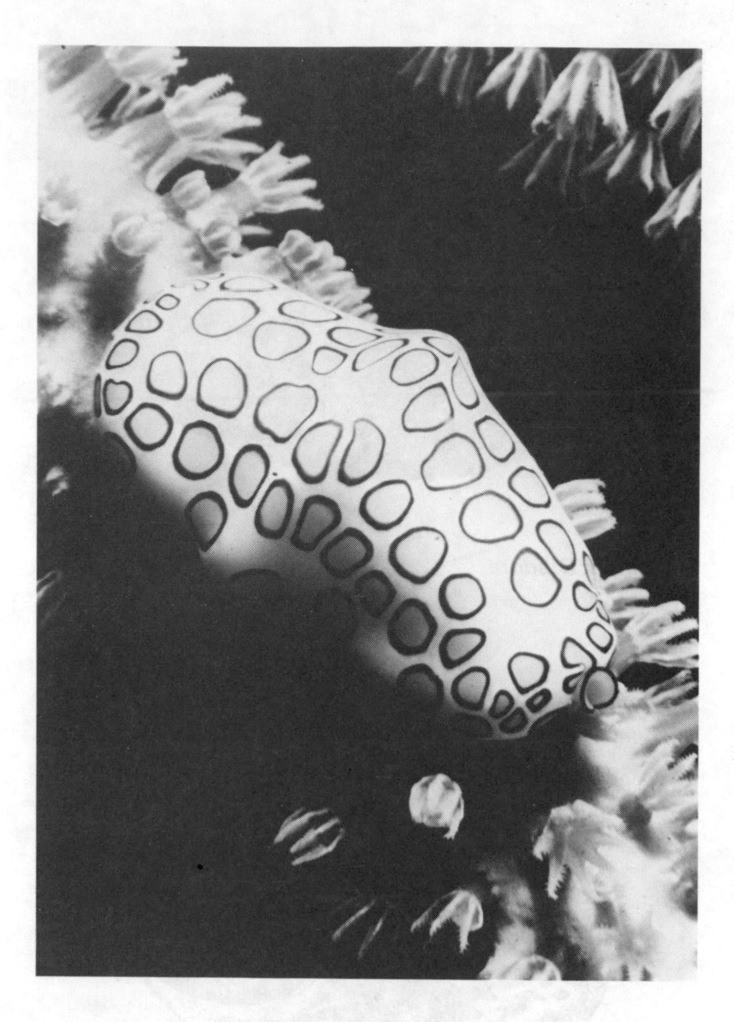

Flamingo tongue snail, *Cyphoma gibbosum*.

Nerites are common and colourful snails of rocky shores. The photograph shows *Nerita tessellata*.

snails are predators of fish and other active animals which are killed by a toxin injected through the specialised radula tooth. Some Pacific species have killed collectors who handled them carelessly.

Bivalve shells are relatively uncommon in the tropical seas and mostly live below the low tide mark. Shells washed up on beaches include the beautiful ribbed venuses and the smooth, colourful tellins. Perhaps the most spectacular bivalve shells come from the mussel-like pen shells, *Pinna*. These transluscent amber coloured shells are triangular in shape and up to 20 centimetres long. In mangrove swamps there are many oysters anchored to the roots.

One need not be a diver to observe the common chiton. These molluscs belong to a group of their own separated from the bivalves and snails and have a shell composed of eight hard plates surrounded by a leathery mantle. They cling tenaciously to rocks in the intertidal zone.

Sometimes the shells of dead snails are used as homes for hermit crabs which as they grow must continually seek larger and larger shells.

The conch and its relatives

There are about fifty recognized species of conch. Around the Caribbean the species *Strombus gigas* is quite abundant and among the largest of the marine snails. Its common name is queen conch. Like other shelled creatures, it eats, reproduces, contains a heart, stomach and a brain. Its stalked eyes peer from the shell curiously as it rests on the shallow, sandy bottom in the flats or near coral formations.

The queen conch crawls about the substrate in true snail fashion, but it is also capable of hopping. The snail raises its shell with its strong, muscular foot and hooked operculum (a hard appendage that acts as a trap door to protect the soft animal when it withdraws inside the shell) and thus lurches across the sand.

The immature conch is called a 'pink roller'. It does not have the full, flared lip of the mature snail and is therefore less stable on the sea bed. This conch is sought-after as food, throughout the Caribbean. Conch steaks, fritters, salads and chowder are among the delicacies derived from this giant snail, and its shell is often used for jewellery. Like the lobster, however, the conch is endangered for the very reason that it is so popular. Unless restrictions are forthcoming that limit the taking of this marine animal, the future does not look bright for its continued existence in large numbers.

The queen conch *Strombus gigas* only develops its wing in maturity. The younger wingless shells are known as rollers. The photographs above show a roller, a young adult with a thin wing and an old adult with a thicker wing.

Other species include the fighting conch with its beautiful deep-orange shell and the rooster tail conch in which the outer lip has a long extension stretching far beyond the top of the spine.

The queen conch has, of course, one of the most spectacular shells in the Caribbean. No less beautiful are the helmet shells, *Cassis*, and the trumpet triton, *Charonia*. The helmets feed on sea urchins and their heavy shells are much in demand by collectors. All of these snails will be in danger of depletion or local extinction unless in some way they are protected from professional and amateur collectors who take them alive to ensure shells in prime condition: a classic case of killing the goose that lays the golden egg.

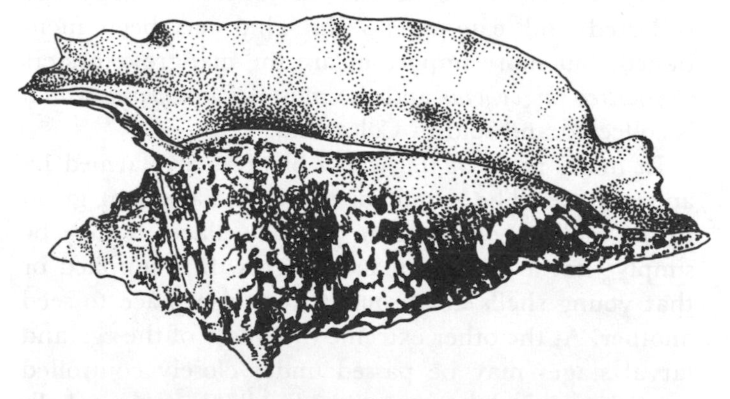

Rooster tail conch, *Strombus gallus*.

Molluscs and food

When a growing animal feeds probably less than a tenth of what it eats is incorporated into its own body, the rest is either excreted or burnt up to supply energy. In the open sea the phytoplankton is eaten by zooplankton which is consumed by larger animals including small fish. These are eaten by bigger fish which are perhaps finally eaten by a 'top' predator. It is these top predators, tuna, kingfish, dolphins and so on that constitute a large proportion of the fish that we catch in the Caribbean. If this fish is the result of perhaps five stages in a food chain and if 90 per cent. of

The king helmet, *Cassis tuberosa*.

the material is lost at each stage then it can be seen that the flesh of the top predator represents a vast amount of phytoplankton. From the point of view of food production it would be much more efficient to either harvest the phytoplankton itself or at least some animal much lower down the food chain.

Many molluscs, especially the bivalves, fit this requirement well. As has been said, bivalves mostly feed by filtering the water in which they live to obtain their food, the filter being composed of modified gills and designed to trap particles of small size, typically phytoplankton and small pieces of detritus. Because of their relatively sedentary nature (only a few can move to any extent) they use relatively little of their food for energy production and their conversion of food to flesh is very efficient. Further, their food is brought to them by the sea; they do not need to move about to collect it and as a result they can live very close together. Given all these features, it is not surprising that the production per hectare of sea bed by bivalves can, in suitable circumstances, far exceed that of cattle being grazed on pastures.

In various places in the Caribbean bivalves are collected and eaten. The chip-chip has been mentioned, but more important are the mangrove oysters *Crassostrea rhizophorae* and the mussel *Purna purna* which is collected and sold in Colombia.

In many parts of the world bivalves are farmed by aquaculture. The extent of the encouragement given to the molluscs varies from place to place. It may be simply that a suitable place for growth is supplied or that young shells are brought from one place to seed another. At the other extreme the whole of the egg and larval stages may be passed under closely controlled conditions in a laboratory which sells the young shells (spat) for seeding.

There are at present a number of attempts being made in the Caribbean to encourage the aquaculture of bivalves. Some of these attempts involve local animals known to be of commercial importance (mangrove oysters). In others it is intended that fast-growing animals such as the oysters *Crassostrea gigas* will be imported from outside the region in the form of spat. There are also investigations being carried out on local animals which are not at present eaten but which may be useful in the future, not necessarily for human consumption but perhaps as an animal feed supplement. Such animals include the small mussel *Brachidontes* which feeds mainly on detritus stirred up in the surf on many beaches. Perhaps it could be persuaded to eat detritus from land plants, e.g. sugar cane leaves.

Common Fish of the Reefs

So great is the number of fish species found in the Caribbean waters that there is scarcely room on these pages to list them all. Some common fish families, their behaviour patterns and their role in the ecosystem, are covered here.

The parrotfishes are herbivores but usually they eat plant life by chewing on hard corals and digesting the encrusting algae and zooxanthellae. They will also eat sea grass if it is available. Parrotfishes are distinctive in that their teeth form a pair of beak-like dental plates. These grind up the algal food with the soft coral rock; the latter is excreted by the fish in the form of sand. In this way the parrotfishes contribute an enormous amount of sediment to the sea bottom, about one tonne of sand per hectare, per year. These fish have many colourful species, one example being the stoplight parrot, the female of which is bright red on the underside and fins whilst the male is predominantly green with three diagonal orange bands on the upper half of the head.

The snappers are one of the most important fish families, being sought after as food. Larger species are found in deep water. Snappers are carnivores, or meat eaters, feeding on crustaceans and small fish. A common inhabitant of shallower reefs is the yellowtail snapper with its broad lateral yellow band, and the mangrove snapper with its faint white vertical stripes.

Hog snapper, *Lachnolaimus maximus*.

The grunts are very similar to the snappers and are usually found in large aggregations on the reefs. The grunts are named from the strange sounds they produce by grinding their upper and lower teeth together. Collecting in small schools by day, the grunts are protected from predators. At night, they disperse to feed individually among the reefs.

The squirrelfish are also nocturnal feeders. They are a very spiny family of fishes with large black eyes and red coloration. During the day, they find a niche or overhang where they remain for protection. Although squirrelfish are seen commonly in shallow waters, there are some species that exist as deep as 100 metres. Squirrelfish are generally good to eat, but are small for use as a commercial fish.

The goatfish are distinct from other inshore fish families in that they possess pairs of long barbels protruding from their chins. Goatfish live in close association with the sand or mud bottom; when they feed, the barbels which have sensory organs on them move rapidly over the bottom and are thrust deep into the sediment. The goatfish feed on small invertebrates beneath the sand.

The surgeonfishes derive their name from sharp spikes at the base of their tails which can ward off attackers. These fish graze on algae: they are either blue (tangs) or brown (doctorfish) in colour, and are often seen in schools.

The groupers which are members of the sea-bass family, are a valuable food fish throughout the islands. The most common is the Nassau grouper, which sports zebra stripes. These fish are carnivores, feeding on fish and crustaceans. Where spearfishing has not been carried on, groupers can be tamed and hand-fed. Although they are usually seen singly, at certain times they come together for mating.

The triggerfish has three dorsal spines which are used to threaten predators. If pursued it seeks shelter in a small recess with a restricted entrance, and raises the first dorsal spine, thus wedging itself firmly in place. In spite of their small mouths, many triggerfish feed on larger, well-armoured invertebrates such as crabs, shellfish and sea urchins. They use their powerful jaws and sharp teeth to break the animals into small pieces. The most beautiful member of this family is the queen trigger with its bright blue markings.

The angelfishes cannot be rivalled for sheer beauty and grace. Deep-bodied and highly compressed, they have slender brush-like teeth, and feed primarily on sponges. Largest are the black angels, sometimes called the greys, and the french angels. Most colourful is

Blue-striped grunt, *Haumulon scirus*.

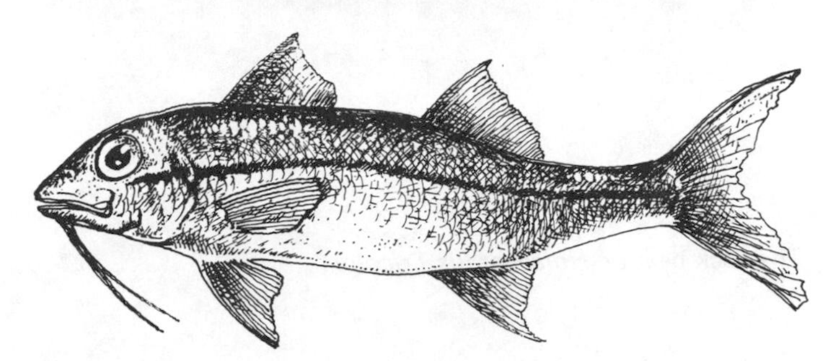

Yellow goatfish, *Upeniens martinicus*.

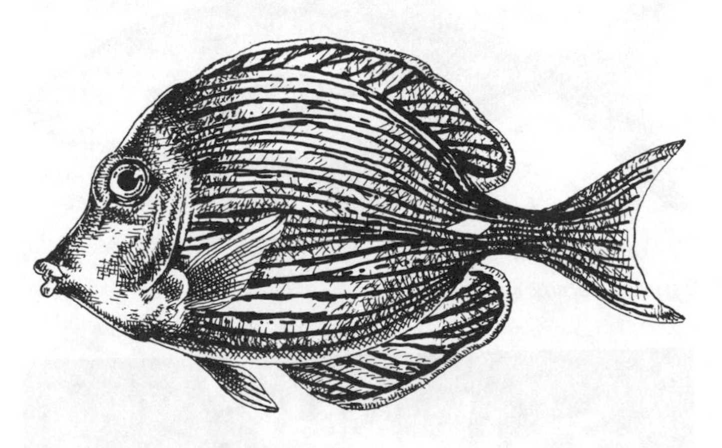

Blue tang, *Acanthurus coeruleus*.

Queen triggerfish, *Balistes vetula*.

The rock beauty, *Holacanthus Tricolor*.

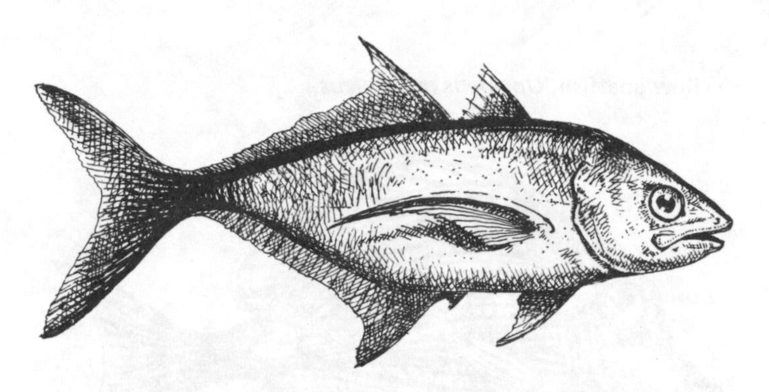

Barjack, *Caranx ruber*.

Porcupine fish, *Diodon holacanthus*.

the shy, blue and yellow queen angel. A close relative, the rock beauty, is striking with its black and yellow body. The french and black angelfishes are tame, and can easily be trained to eat from a diver's hand.

The butterflyfishes are closely related to the angelfishes. They are small in size but also deep bodied and are usually seen in pairs, grazing along the reefs.

The jacks are a strong fast moving family, carnivorous and silver in colour. They are sleek, and have scimitar-like tails. Their many species include the great amberjack, the barjack, crevalle jack, pompano, palometa, and the permit. These fish are usually seen in transit from one point to another.

The wrasses are a truly large family of fish found on the Caribbean reefs and perhaps the most diversified of all fish families in body form and size. The largest is the so-called hog 'snapper'. However most wrasses are small, swarming about on the shallower reefs. They feed upon invertebrates, and are diurnal. At night many smaller species bury themselves in the sand. They are usually brightly, and often gaudily coloured.

The porcupines are known for their ability to expand to several times their normal size, when pursued. The body is covered with spines that protrude when the fish swells, thus discouraging a predator. The fish has a large, powerful, beak-like jaw to crush hard-shelled invertebrates upon which it feeds.

The flounders, or flatfishes, are distinctive in that both eyes are on one side. In the larval stage the eyes are in a normal position, until one begins to migrate to the opposite side of the head. A colouful species is the peacock flounder, which is hard to find when resting on the sandy bottom but as soon as it starts to cruise over the sand distinct blue markings appear.

Fish species of Caribbean reefs are amazingly diverse, each having its own place in the reef ecosystem. A significant depletion of one family of fish can effect the health of the reef as a whole.

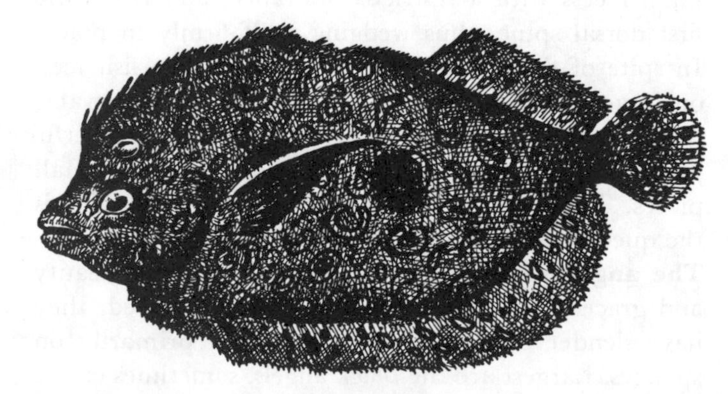

The peacock flounder, *Bothus lunatus*.

Misunderstood 'Monsters' of the sea

It is regrettable that legends, fictional stories and movies have caused man to fear some of the sea's most harmless creatures.

The stingrays are very primitive fish with cartilaginous skeletons. They usually lie on the bottom, partially buried in the sand, and feed by excavating shallow depressions to expose invertebrates. Many rays bear their young alive. The majestic spotted eagle rays attain a width of about two metres. They glide singly or in small groups over shallow sand flats or deep reefs. Very shy, rays will usually flee from any diver; if stepped on, however, a ray will use his sharp barb, at the base of the tail, for defence. If stingrays are approached from the side or the front they can raise their sting making thrusting movements. Approached from behind they are unable to fence in this way and are relatively innocuous.

The sharks, relatives of the rays, are abundant in all seas, and are perhaps most feared by man. Some are harmless; others are not. Their primitive nature and reputation as voracious scavengers make divers cautious in their presence. Close to shore sharks are seldom seen but an occasional nurse shark may be encountered sleeping on the bottom. It feeds on small invertebrates and is shy by nature, but still capable of biting!

The octopus, a highly intelligent mollusc, attains a length of only about 40 centimetres in the Caribbean. By day it hides in a hole on the bottom, discernible by a pile of empty shells at the entrance, leftovers from its meals. It can change colour rapidly, to

The spotted eagle ray, *Stoasodon narinari* can be seen cruising the reefs at all depths.

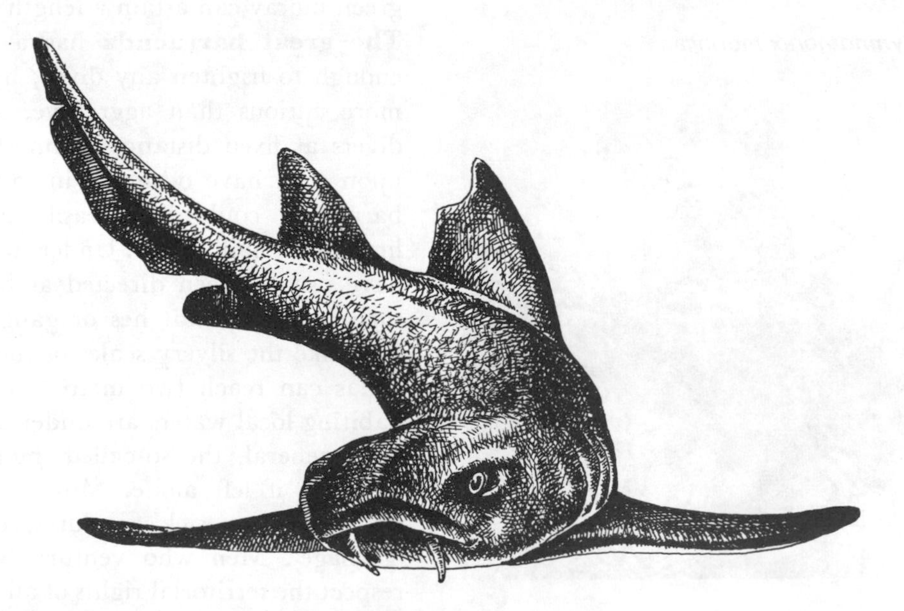

Nurse shark, *Ginglymostoma cirratum*.

Caribbean octopus.

blend with the surroundings, as can its relative the squid, also found in local waters and growing no larger than about 30 centimetres. Both creatures emit clouds of black ink when frightened, a defence mechanism that hides them from an attacker.

The eels occupy their coral caves, or remain buried in the bottom sediment, by day and search for food at night. Normally shy, eels have been known to attack spearfishermen carrying dead or dying fish. Local species include the congers, snakes, and morays. The green moray can attain a length of two metres.

Spotted moray eel, *Gymnothorax moringa*.

The great barracuda has a nasty sneer which is enough to frighten any diver; however, these fish are more curious than aggressive, and habitually follow divers at fixed distances. Almost all recorded attacks upon men have occurred in murky water where the barracuda could have easily mistaken the bather's limbs for a flailing fish. Under such conditions the bites have usually been directed at bright metallic objects such as divers' watches or gauges which presumably flash like the silvery scales of the normal prey. Barracudas can reach two metres in length, but most inhabiting local waters are under one and a half metres.

In general, the so-called 'monsters' of the sea are harmless if left alone. Most are scavengers, feeding upon dead or sick creatures, cleaning up Nature's 'garbage'. Men who venture beneath the sea must respect the territorial rights of all creatures found there, for it is their world, and Man is only an uninvited guest.

The great barracuda, *Sphyraena barracuda*, is usually harmless.

The trumpetfish, *Aulostomus maculatus*, tries to hide itself against a travelling tiger grouper, *Mycteroperca tigris*.

Mechanisms for offence and defence

It has been mentioned that fish school for protection, for there is indeed, safety in numbers. Some fish however, use rather artful forms of camouflage, both to capture and to hide from other creatures. An example is the trumpetfish, long and slender. It may hang vertically in the water next to a tall coral branch, attempting to disguise itself. Occasionally the clever trumpet may travel with a school of tangs, appearing appropriately blue, to blend with its companions. At other times the trumpets are seen literally riding the backs of groupers in an attempt to be 'invisible'.

Some fish are endowed by Nature with disruptive coloration, intricate markings on the body of the fish that confuse a predator. The four-eye butterfly, for instance, has a large black spot at the base of its tail. Resembling an eye, it fools the pursuer into miscalculating the direction of the chase.

Some fish contain toxins in their flesh or exude poisonous mucus. This is the case with the trunkfishes. It is presumed that these poisons are defensive. Poisonous or harmful animals often advertise their presence with bright colours (e.g. yellow and black wasps, coral snakes and many distasteful insects). However, this does not often seem to be the case with fish, perhaps because many quite harmless fish are brightly coloured for other reasons. Indeed some of the most hazardous fishes in the Caribbean, the scorpionfish, are very well camouflaged (although it must be admitted that some

Four-eye butterflyfish, *Chaetodon capistratus*.

Scorpionfish, *Scorpaena plumieri*.

Indo-Pacific species are brightly coloured). These fish, sometimes called stonefish, are aptly named for they do indeed look like stones lying on the sea bed or on a coral reef and they do have a sting like a scorpion. The poison is contained in sacs at the base of the dorsal fin spines. An unwary predator or human will find these poison barbs very unpleasant indeed.

The toxic fish mentioned above produce their own poison. However, occasionally, fish of various types can cause poisoning when eaten. These are usually predatory reef fish (e.g. barracuda) and the poison they contain has been passed to them from the fish upon which they feed. These in turn have received the toxin from their food and so on. The original source of the poison is in an alga. Research is being carried out to devise a simple test to detect the poison in affected fish.

Nature has endowed each reef creature with some means of self-defence, as well as an effective method of finding food. In this way, fish populations can remain relatively stable, as long as man does not interfere.

Fish colours and territories

Why it is that many reef fish are so colourful? There may be different answers and certainly there is no one reason which applies to all the colourful species. However, many are undoubtedly advertising their presence to other members of the same species. Such fish live relatively solitary lives, not forming schools but staying put on one small piece of reef. If another member of its own species wanders by it will be chased out of the area which the owner fish regards as its territory. Usually these encounters do not result in injury because the intruder seems to accept its misbehaviour and loses no time in rushing away. If however the intruder cannot easily escape then fights will result which can, and do, result in the death of one or other of the combatants. It is for this reason that only one individual of such species can be kept in a small aquarium tank. Sometimes other species will be attacked if they have either similar colours or a similar shape.

Territorial behaviour is commonly found in a wide variety of animals including birds and mammals. The territory may be held for a variety of reasons; to preserve a large enough area for feeding, to protect a breeding site or to attract females. In such circumstances the territory holders will inevitably spend much time fighting off intruders and bright colours help to avoid prolonged intrusion deep into territories as would be the case if the animals were unobtrusively coloured.

These bright colours carry some disadvantages however, for they render their carriers obvious to predators and one often finds that such fish are adapted for retreating into holes or crevices in the coral in the event of a predator appearing. Another problem is that if the bright colours of one's relatives provoke aggression how can mating take place. This is solved in a variety of ways. Often, for example, the adults are not so brilliantly coloured, and not so provocative to one another. A good example is the french angelfish which is striped black and gold in the juvenile but dark greyish in adulthood. The yellow-tailed damsel fish when young is bright blue and covered with lighter

The juvenile french angelfish, *Pomacanthus paru*, is a brilliant yellow and black fish. As it gets older its stripes fade and are replaced by light-tipped scales over the whole body surface. The top picture shows a juvenile, while in the middle is an intermediate form and at the bottom is a fully grown adult.

The sergeant major, *Abudefduf saxatilis*, is a strongly striped fish. However at certain times these stripes become much less obvious and thus less provoking to other fish.

iridescent blue spots but in the adult form is much less striking. Some normally bright fishes take on much less bright colours when mating. The yellow, black and silver striped sergeant majors become a nearly uniform dark blue when guarding the eggs. Other brightly coloured fish lose their colours when asleep.

One species of fish, the wrasse blenny, looks very like the juvenile stages of the bluehead wrasse. The bluehead is a cleaner fish and is therefore not taken by predators. Because of its resemblance the blenny seems to be able to go abroad on the reef to get its food, small fish and crustaceans, without being attacked by larger fish. This mimicry goes beyond coloration and body shape even to an imitation of the bluehead's way of swimming.

Fish and the reef

Fish on the reef have an almost infinite variety of life styles. Between them they are able to consume almost any plant or animal material available. Only a select few animals are immune from attack by fish. Most of these are either distasteful or perhaps well protected like the spiny black sea-urchin, *Diadema*, that is so common on the reef. Its very commoness is a testament to its ability to avoid predation. There is no doubt that it makes good eating once the predator can get past the spines, for if one is broken open clouds of fish, especially the small wrasse known as slippery dicks, swim around to pick the flesh from inside the shell.

Many fish are predators on other fish. The barracuda relies on its speed over short distance to catch its prey while others lie disguised against their background waiting for unwary passers-by. The scorpionfish fall into this group, as does the curious batfish. These latter have a small fleshy protuberance sticking out from the front of their heads. By gently jiggling this they attract small fish within reach of their capacious jaws. Batfish when young, are also said to have an uncommon resemblance to fallen mangrove leaves as they lie on the sea bed. Perhaps this disguise protects them from predators as well as hiding them from their own prey. As has been mentioned, trumpetfish make themselves unobtrusive by colour change and skillfully placing themselves by long slender objects.

Parrotfish are grazers. If they can get it they will eat algae or sea grass but often these are unavailable and the parrotfish will use its powerful jaws to rasp the surface off living coral heads. They favour the large smooth colonies of corals such as *Siderastrea*. It is not uncommon when snorkelling or diving to see the marks left by their jaws on the surface of the coral and perhaps to even hear them rasping away at their work. The fish obviously remove quite large quantities of coral skeleton with the nutritious part of their meal and thus their faeces contain much fine ground coral sand. This sand settles into the spaces in the reef where eventually it is consolidated. This activity of the parrotfish is of considerable importance in reef-building.

Many fish live on crustacea and molluscs in the sand near to reefs. Among these are the goatfishes which can be seen in schools moving across the sea bed periodically pushing their heads into the sand and alternatively sucking and blowing water through their mouths. This action throws up the sand and creates a shallow hole. In this hole the goatfish searches for food with the long barbels which are thrust down from its lower jaw.

The black and white spotted drum, *Equetus punctatus*, is a shy but beautiful reef dweller.

The shape of fish may open up certain areas for exploitation. For example, the handsome gold-spotted eel can slither its snake-like body into all sorts of crevices and burrows in the search for food.

Many reef fish are far more active at night than during the day. This makes a night time trip with an underwater torch particularly interesting. Among these nocturnal fish are the squirrelfish, the voracious moray and, unfortunately, many species of shark. To guard against being snapped up in their sleep by these night time prowlers many fish hide away in crevices and bury themselves in the sand (e.g. the wrasses). Some parrotfish secrete a layer of mucus around themselves while they sleep which presumably discourages would-be predators.

Fish watching

One of the glories of the reef is undoubtedly its fish. The variety of colour, shape and movement is staggering and to scuba dive over such reefs is like swimming inside an aquarium. But does one need to go to the lengths and expense of aqua-lung diving to enjoy these sights? The answer is clearly 'no'; there are many ways in which one can observe beautiful fish. The simplest is to go to the fish wharves and markets where the fish, sometimes still alive, can be easily examined and one can learn much about their habits and so on from the fishermen too. The major problem is of course that these are literally fish out of water and after death their movements, and often their colours are lost.

Why can one not walk out and watch the fish through the surface of the water? In some places one can, but even the slightest breeze or wave will disturb the surface making a clear view impossible. However, this can be counteracted by using a glass bottomed viewing box. This can be simply made, being a wooden or metal box about 30 centimetres square and perhaps 40 or 50 centimetres deep. One end is open and the other covered with a piece of fairly thick glass and the whole made watertight with putty. The observer holds the box in the water so that the glass bottom is below the surface and looks through this glass-window from the top of the box. The same principle is employed in the glass-bottomed boats so common in tourist-developed

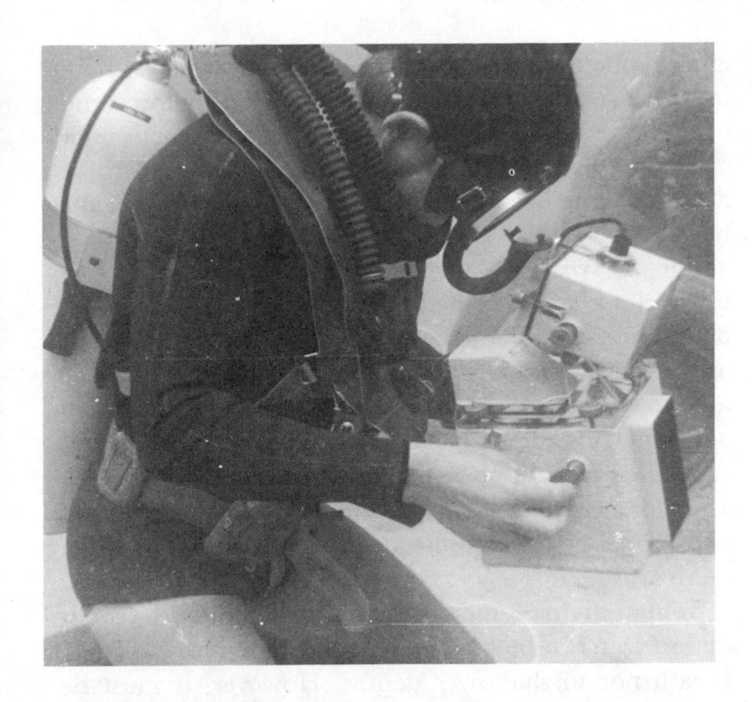

Better to catch a fish on film than with a spear! Here an underwater photographer uses a twin lens reflex in a waterproof housing.

Artificial reefs are a convenient way of getting rid of rubbish whilst encouraging fish populations. This one is made of old motor tyres roped together.

The fish photographer takes to the depths complete with a twin lens reflex in a waterproof housing and a flash gun.

Who is watching who? A one metre long trumpet fish, *Aulostomus maculatus*, eyes a diver.

A group of sinister looking, but harmless, lizardfish, *Synodus intermedius*, pose for the photographer.

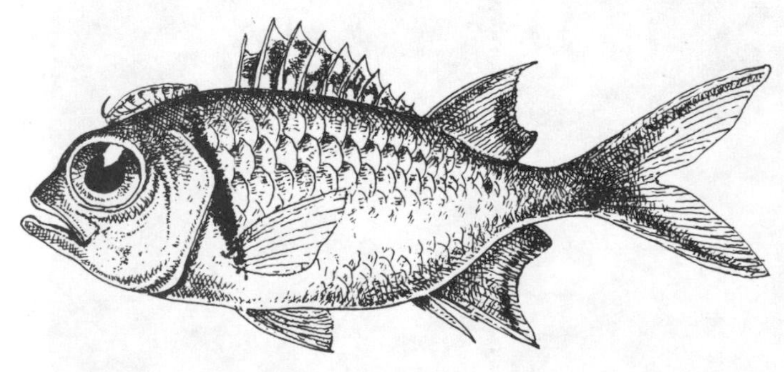

The blackbar soldierfish, *Myripristis jacobus*, wears a parasitic isopod on its forehead.

areas. Here the window is somewhat bigger and is built into the bottom of a boat which can be motored out over the reef while its occupants get an excellent view of the sea bed. In clear waters it is not unusual to be able to see the bottom in 30 metres of water from such a boat.

For the swimmer a mask and snorkel have the visibility of a glass-bottomed box but with added freedom of movement. The snorkeller may prefer to float at the surface but if he takes to diving down he will be able to see the fish from the side as other fish see them and not just their backs from above. Experienced snorkellers can of course dive to considerable depths, 15 metres or more is not uncommon, but the beginner will probably be contented with dives of two or three metres.

Scuba diving gives one almost complete freedom under water. One is no longer limited to a single breath nor to shallower depths. However, it must be remembered that this type of diving has many potential dangers and properly supervised training is essential. This training is sometimes available at hotels but on many islands there are non-profit making clubs where for a relatively small sum very good training can be had, though it may take some weeks to complete.

Many countries in the region are also developing public aquaria where visitors and residents alike can go and study the local marine creatures under dry and convenient conditions.

A number of excellent publications exist to help the watcher identify his fish, some of these are listed in the bibliography at the back of this book.

Reef Togetherness

A coral reef is a crowded community where diverse creatures live 'elbow to elbow'. While many organisms are continuously preying upon others for food, there exist several relationships, pairings of different species, for the benefit of one or of both. These are closer and often more than specific prey/predator relationships. Four types will be mentioned.

Parasitism occurs when one organism lives in or on the body of another, obtaining nourishment at the host's expense. An example found on coral reefs is that of the soldierfish and a small crustacean called an isopod, that fastens to the fish's forehead. Butterflyfish and grunts may also be seen to 'wear' isopods between or below their eyes. Because they cannot reach their bodies with their mouths and lack limbs for scratching, fish are plagued with these so-called ecto-parasites, many of which are too tiny to be seen but nevertheless cling stubbornly to the helpless fish.

Mutualism is an association which benefits both parties and to some extent offsets the one-sided parasitic relationship. Fish with small body parasites visit 'cleaning stations' throughout the reef. Waiting there are tiny cleaners, which may be fish or shrimp, who spend their day picking the parasites from the bodies of the patient and grateful clients. In some cases the cleaners will venture into the mouth or gill cavities of their customers. Most cleaners advertise their presence by their coloration and/or particular dance they do. Shrimp, for example, wave their long antennae frantically when a large fish comes near, broadcasting the fact that they are cleaners. In one scientific experiment, cleaners were removed from a small patch reef; soon the larger fish, overburdened by parasites, sickened and died.

Commensalism, meaning common table, is an association in which the benefit may be one-sided, although unlike parasitism the other animal appears unaffected either way. For example, one not infrequently finds polychaete worms living with hermit crabs in their shells. When the hermit crab is eating, the worm will appear and scavenge discarded food. This relationship clearly benefits the worm who gains protection and free food while apparently doing his host no harm. Likewise, there is a small fish which lives in the rectum of large sea cucumbers. It uses its strange home for protection, emerging to feed and on returning reenters tail first. The sea cucumber is apparently unharmed by this strange visitor. However, the line between commensalism and parasitism is a very finely drawn one. This is the case with the brittle star, *Ophiothrix*, which lives attached to sponges and soft corals holding out its arms to filter the water. It is possible, perhaps, that the brittle star is depriving its

A tiny cleaner fish ridding a moray of its ectoparasites. Note the striking colourisation of the cleaner which acts as an advertisement.

host of food by getting first bite. One commensal that does seem definitely to damage its host is the tiny pea crab that lives inside the mussel *Brachidontes*. In this case there is clear evidence that the mussel's gills are damaged by the crab. Curiously the same relationship exists with the European mussel and in this case it is difficult to show any deliterious effect.

There is one beautiful example of an animal which has a commensal relationship with one creature and a mutualist relationship with another. This is the tiny shrimp *Pereclimenes*, a particular species of which teams up with an anemone, and has a perfect refuge within the stinging tentacles, being obviously immune to them. This shrimp cleans fish as well, emerging from its tentacled refuge to pick parasites from passing clients.

Symbiotic relationships are really specialised forms of mutualism in that they favour both partners, but they are often so close and intimate that the pair become interdependent and cannot readily survive in isolation. The coral itself is the best example here, for as has been explained, many of the coral polyp cells contain unicellular algae, each contributing to the other's well being. Many other cases of symbiosis are known often involving organisms from very different groups. Hydroids or sea-anemones often attach themselves to the shells of hermit crabs while other crabs have sponges growing on their shells. The advantage to the crabs lies in protection or camouflage while the advantage to the sessile creatures is that of locomotion. Another marine example is that of the yellowish lichens that can be seen on some rock surfaces in the splash zone. These, like all other lichens, are composite creatures containing both an alga and a fungus. So intimate is this relationship that the resultant plant can in many ways be deemed to be new and independent more than simply the sum of its parts.

The presence of so many examples of mutualism and symbiosis in the coral reef is one indication of the extreme closeness of the community and, through the retention and efficient transfer of nutrients one of the factors which leads to its success.

The anemone, *Condylactis gigantia*, harbours the tiny shrimp *Pereclimenes yucatanicus* in a lifetime partnership.

The whole of this ecosystem rests on the symbiotic relationship between the coral, and its zooxanthellae.

The Coral Ecosystem

A living coral reef is a community composed of thousands of different members living in harmony with one another. The existence of the reef is based upon physical, biological and chemical interactions among all its inhabitants. This interdependence is so vital that many of the reef dwellers cannot live outside the reef zone.

Sunlight, water, fish and lower animals play their role in building and sustaining the coral reef. Sunlight fosters photosynthesis necessary to the tiny single-celled plants that live in the coral tissues. Water brings nutrients to the entire community. The lower animals secrete hard calcium carbonate that cements the community together, and the excrement of fish and other animals helps to build up the reefs.

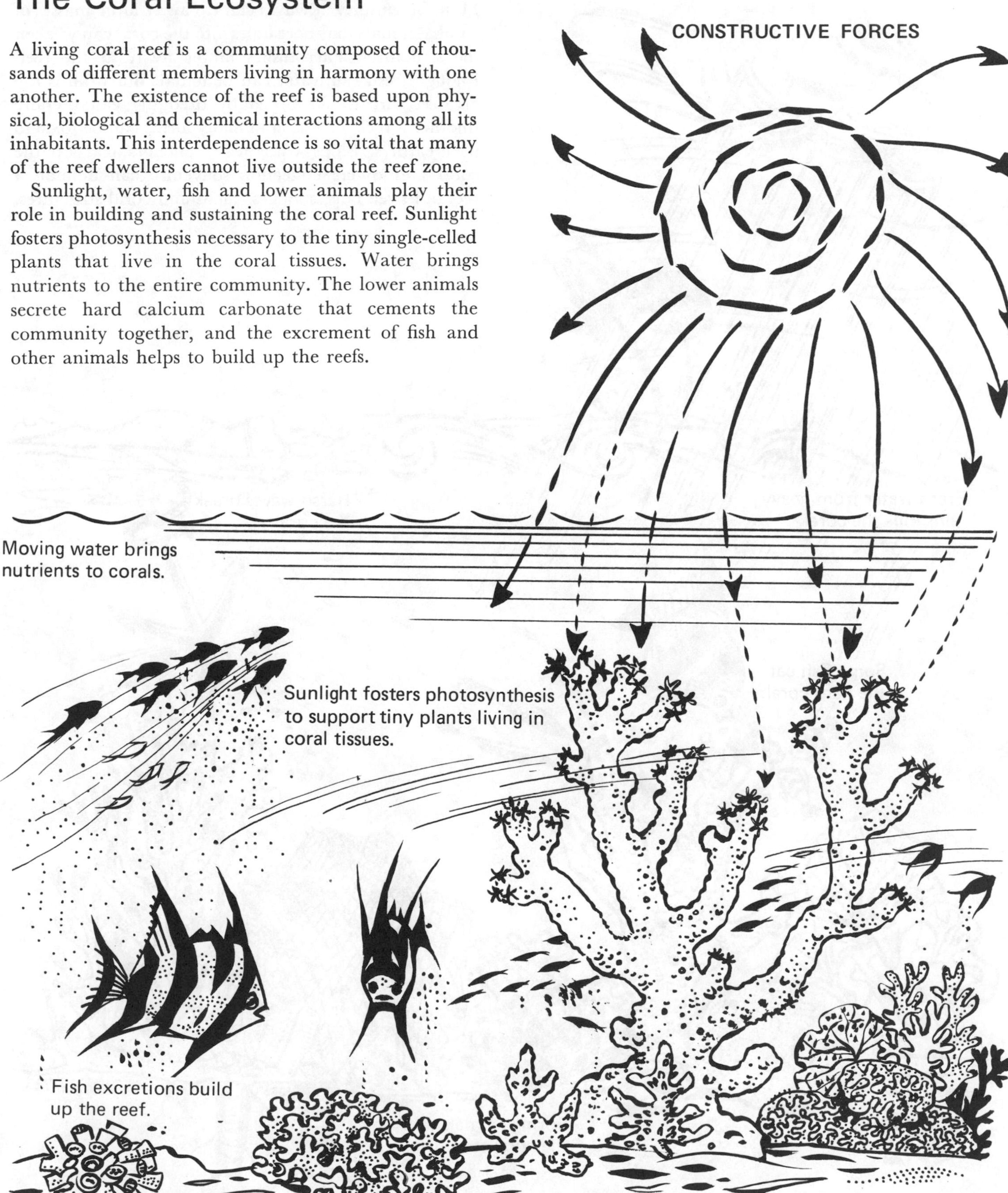

DESTRUCTIVE FORCES

Destructive forces are created both by Nature and by Man. Storm-driven waves can rip away large masses of coral. Animals that bore holes into the coral can weaken the structures. Parrotfishes nibble away at the reef. Enough fresh rainwater can be lethal. But Man is the greatest threat. Silt raised by dredging can literally smother a reef. Pollution in many forms can be toxic to its creatures. Ships with heavy anchors tear up the coral and divers collecting souvenirs can denude a reef of its sea fans, sponges, shells and coral structures.

The Open Sea

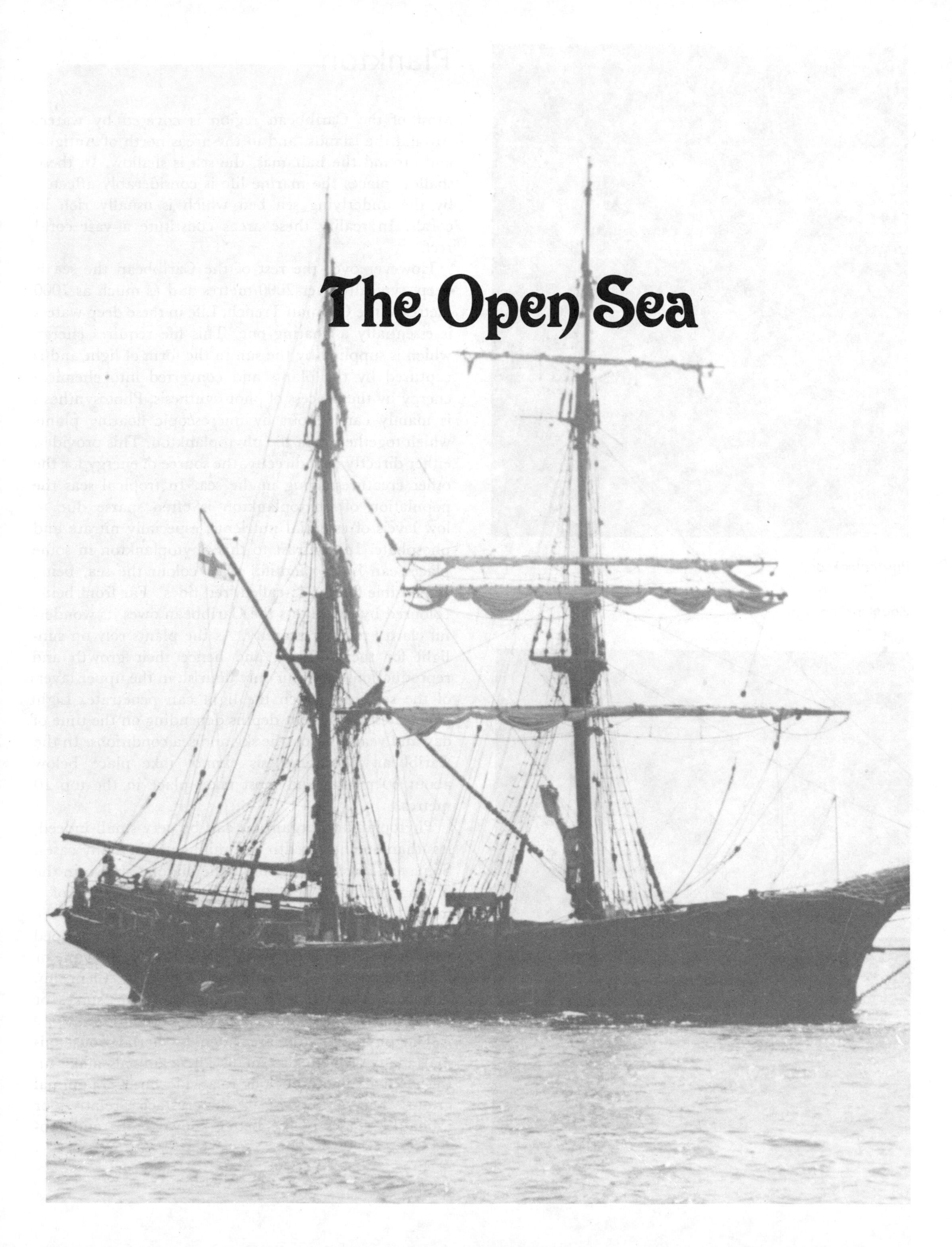

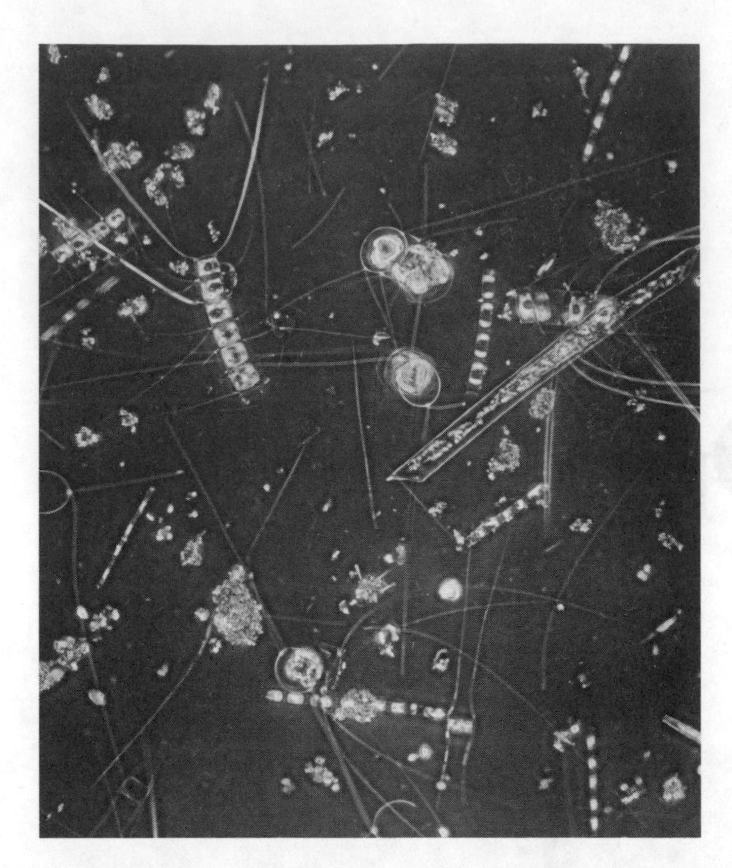

Phytoplankton.

Zooplankton.

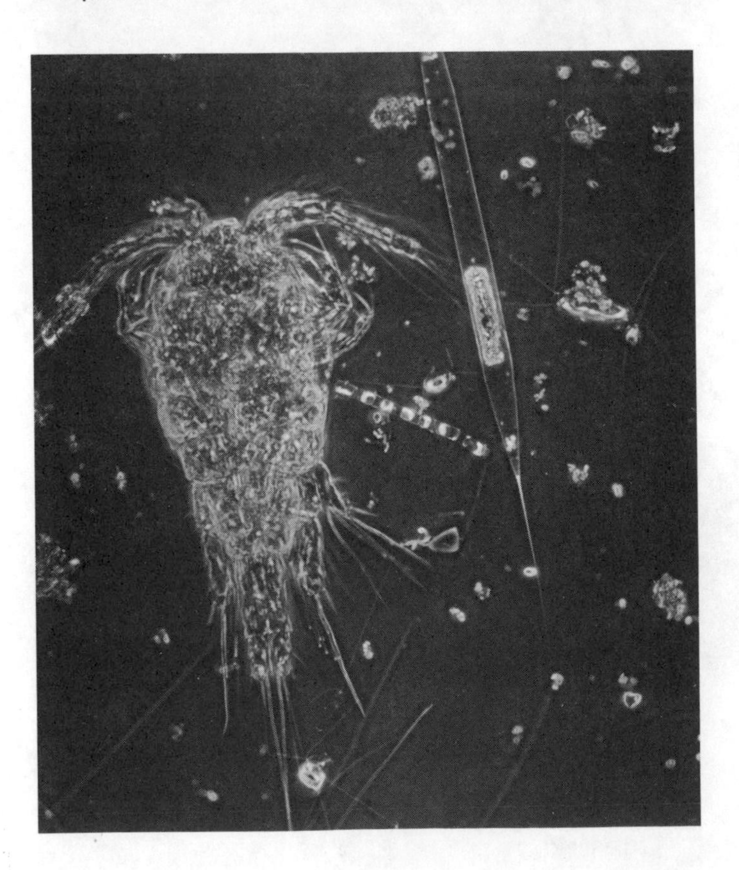

Plankton

Most of the Caribbean region is covered by water. Around the islands, and in the areas north of Antigua and around the Bahamas, the sea is shallow. In these shallow places the marine life is considerably affected by the underlying sea bed which is usually rich in corals. In reality these areas constitute a vast coral reef.

However over the rest of the Caribbean the sea is deep, typically over 2000 metres and as much as 7000 metres in the Cayman Trench. Life in these deep waters is essentially a floating one. This life requires energy which is supplied by the sun in the form of light and is captured by the plants and converted into chemical energy by the process of photosynthesis. Photosynthesis is mainly carried out by microscopic floating plants which together form the phytoplankton. This provides, either directly or indirectly, the source of energy for the other creatures living in the sea. In tropical seas the population of phytoplankton is often sparse due to low levels of essential nutrients, especially nitrate and phosphate. In contrast to this phytoplankton in some places can be so plentiful as to colour the sea, being responsible for the so-called 'red tides'. Far from being coloured by the plants the Caribbean owes its wonderful clarity to their scarcity. As the plants rely on sunlight for their energy, and hence their growth and reproduction, they can only flourish in the upper layers of the sea into which the light can penetrate. Light penetrates to different depths depending on the time of day and year, and on the sky and sea conditions. In the Caribbean photosynthesis cannot take place below about 60 metres and most takes place in the top 10 metres.

Phytoplankton organisms can be very small indeed, less than one hundredth of a millimetre in many cases. Some are larger, for example the diatoms shown in the photograph may be anything up to one tenth of a millimetre. To help prevent them from sinking into the dark zone of the sea these tiny plants are often equipped with hair-like flagella with which they can swim, or lighter-than-water oil droplets that buoy them up. Others slow down their sinking by having spikes or spines that act in a similar way to a parachute.

The phytoplankton are eaten by herbivorous animals many of them also small. These animals make up the zooplankton and belong to a wide variety of animal groups. The commonest are the crustacea, relatives or immature stages of the better known crabs and lobsters. Usually they are barely visible to the naked eye (about half a millimetre). Some may pass their whole life in the plankton while others are young or larval stages of animals which, when adult, live elsewhere. Many bottom-living crabs, molluscs, worms and echinoderms start their lives by spending weeks or months drifting in the upper layers of the sea. These small larvae are able to exploit the tiny phytoplankton plants until they are large enough to cope with bigger food. These swimming and floating youngsters also help to distribute their kind over a large area, a fact of particular importance to sedentary or attached species. It is known, for example, that the larvae of the pen shell, *Pinna*, can drift the full width of the Atlantic to settle on the east coast of the USA having been spawned on the coasts of Africa.

Larger animals also live out their lives in the deep seas, some in the illuminated upper layers, some in the dark abyssal depths. In the surface regions many of these animals are filter feeders (see p. 23) sieving out the plant and animal plankton. The largest animal in the world, the blue whale, is one such, straining thousands of litres of sea water each day to extract its crustacean food. Most, however, are much smaller and include the larvaceans, floating relatives of the sea-squirts (see p. 16). Others catch their prey with stinging tentacles. One of these is the beautiful but unpleasant portuguese man-o'-war, *Physalia*, a relative of the jellyfish and sea-anemones. This animal has a gas-filled bladder which floats at the surface of the sea. It is coloured brilliant blue or violet and may be 15 centimetres in length. From the bladder tentacles several metres long trail down into the sea armed with powerful stinging cells. Large specimens catch and eat quite big fish. When floating near beaches or washed up on the sand they are a considerable hazard to bathers and should be avoided if at all possible. The sting is very painful at best and at worst a danger to life.

Floating snails are also attractive creatures and altogether safer to handle. There are two main groups, one which swims by means of an enlarged foot and the other which hangs in a floating mass of mucus and air bubbles. *Janthina* is an example of such a pelagic snail; its beautiful violet and purple shell is occasionally washed up on beaches.

There is only one large oceanic plant of any importance; the sargassum weed. This floating seaweed is a member of the genus *Sargassum*, many species of which are normally attached forms found on rocky shores. The floating forms are capable of completing their whole life cycle far from land. This seaweed is common in the Caribbean and has an associated fauna of animals adapted for life in its forest. These include bryozoans, small colonial creatures, the crab *Portunus sayi* and a number of fishes. Most of the inhabitants are camouflaged which makes them difficult to detect. The sargassum fish in particular have taken the form and colour of the weed and are very bizarre creatures. The inhabitants of the weed are not typical pelagic animals but rather related to the inshore animals from which they have evolved.

Sargassum sp is often found as a pelagic floating weed. However it is also a common attached weed. It is shown here on a rocky shore. Note the reproductive fronds. When it is in its floating form it develops buoyant bladders.

The animals that live deep in the dark regions depend, either directly or indirectly, on the animal and plant material that 'falls' from the productive layers above. Some of course survive by eating their fellows whom they often lure with a luminescent 'bait' which they dangle just in front of their mouths like the weird batfish. Although much of what falls from above is snapped up a small amount settles on the sea bed where it is eaten by bottom dwellers (including many brittle stars and bivalves) or broken down by bacteria. Needless to say the quantity of food that gets down that far is small and this, combined with very low temperatures, means that growth is slow. Evidence is just beginning to appear which suggests that some of the tiny deep sea bivalves may be over 100 years old!

The hard shells of planktonic organisms are not all recycled and may accumulate on the sea floor to enormous depths, perhaps hundreds of metres thick. Many present day limestones, chalks and silicious earths are the result of such accumulations from the past.

Much of the fish caught in the Caribbean is sold locally in small fish markets. Here we see buyers and sellers awaiting the arrival of the fishing boats in Bridgetown, Barbados.

Fish

At the very top of the food-chain of deep sea herbivores and carnivores come the large pelagic fish. These powerful, streamlined fish are almost always good eating and this, combined with their violent activity when hooked, has made them favourites with professional and sport fishermen alike. The encounter between man and these huge fish has been immortalised in Hemingway's *The Old Man and the Sea*: 'Never have I seen a greater, or more beautiful, or a calmer or more noble thing than you, brother' says his hero of a 400 kilogramme marlin.

Commercial fishermen are usually less ambitious about their fish and a 10 kilogramme kingfish or dolphin, as well as smaller tuna and albacore, are more typical catches. These are most often caught on lines trailed behind the boat, but are sometimes netted. Netting is usually carried out by very large fishing boats which may travel many thousands of kilometres to find good fishing grounds.

Marlin are all fast swimming predators and include some of the most famous sport fish. Many of them have the bones of the upper jaw extended into a long, pointed rostrum. It is said that this 'sword' is used to club smaller fish. There are also numerous records of ships and boats being attacked by swordfish. The penetrative power of the sword would be hard to believe if it were not for many examples of fish breaking them off and leaving them embedded in the boat's woodwork. The British Museum has such a sword that had penetrated 55 centimetres of planking before snapping.

Tuna are not usually as large as the marlins and often hunt in schools. Fishermen who can locate these schools sometimes catch hundreds of kilogrammes of fish in a few minutes as they will take baited hooks with no hesitation. They are very fast swimmers and, although authenticated reports are difficult to obtain, speeds of up to 55 kmph are probably not an underestimate. Three factors contribute to their speed. Firstly, they are beautifully streamlined, fitting almost exactly the engineer's specifications for maximum hydrodynamic efficiency. Secondly, their tails are strengthened by the continuation of the backbone into it, unlike other fish whose tails contain only small bony rays for support. As a result the tail of the tuna is remarkably stiff and strong. Thirdly, many tuna have been shown to have a body temperature higher than the surrounding sea by as much as 4-5°C thus making them warm-blooded to some extent. This raised temperature enables them to work faster than their cold-blooded relatives.

Dolphins are prized for food throughout the Caribbean. The name often misleads visitors who are used to applying it to small mammals related to the whales. These latter are of course warm-blooded and suckle their young in the typical mammalian way while the Caribbean dolphin is very much a fish. The name may have been used because of the high crest of bone on the head which gives the creature a superficial resemblance to the small whales.

The dolphin is a long, rather narrow fish which may be nearly two metres in length and weigh 20 kilogrammes or more, although smaller specimens are more usually seen in the fish markets. In life it is brilliantly coloured, bluish above and yellow beneath but these colours quickly fade after death.

Flying fish are common in all tropical seas but unlike the species mentioned previously they are relatively small, rarely exceeding 45 centimetres in length. The Caribbean species is about 25 centimetres long and rather the shape and colour of a herring. Of course the most remarkable thing about these fish is their ability to 'fly'. For many years there was debate as to whether the fish actively fly or just glide, but high-speed cine-films have now shown that they only perform the latter. Just prior to take-off the fish swims rapidly towards the surface. It bursts out into the air at a velocity of something like 55 kmph and once it is

Two fishermen bring ashore part of their days catch: a dolphin and a sack of flying fish.

Male dolphins have a high crested head.

Flying fish showing the very large pectoral and pelvic fins which act as 'wings' when the fish is gliding through the air.

At present much fishing is done is small boats like these moored off the east coast of Barbados.

airborne it spreads out its large winglike pectoral fins. Some species, including the common Caribbean one, also extend their pelvic fins and so appear to have four wings. The fish then skims over the surface of the sea, usually for a few seconds although flights of up to 15 seconds have been recorded. The distance travelled may bc as much as 350 metres but is most often just 30-40 metres. This ability to 'fly' must help the fish escape from their predators, though there is some evidence that they do it 'just for fun'.

Although these fish are fairly common in the Caribbean they are only fished in any numbers in Barbados and Grenada. The season extends roughly through the first half of the year and the method of catching them is interesting. They are mainly caught at the time of spawning when the females are searching for floating objects on which to lay their sticky eggs. The fishermen go out in their boats and throw palm fronds and other debris into the sea which, together with a small basket of rotting fish held over the side, attracts the flying fish. They can then be caught in small dip nets and lifted into the boat. In recent years these traditional methods have to some extent been replaced by the use of gill nets.

Sharks: no section on fish is complete without mention of these notorious creatures. They have such a reputation and are surrounded by so much mythology that it is difficult to arrive at a balanced view about them. There can be no doubt that most species of shark are voracious predators and some may make unprovoked attacks on swimmers and divers. It is possible that at least some shark attacks are in defence of territory rather than as a feeding response but this has not been clearly established. Luckily the Caribbean does not suffer from shark incidents to the same extent as the Pacific. The number of authenticated attacks is small but nevertheless the sharks are there and should be treated with respect.

The sharks which are potentially the most dangerous are those that feed on large, active prey and include the mako, tiger and hammerhead. However even the usually placid nurse sharks have been known to bite if annoyed. Spear fishers are well advised to trail their catch on a long line well away from their bodies or to place it in a boat immediately after capture for sharks are very sensitive both to the movements of injured fish and the smell of blood.

Sharks are caught and eaten throughout the Caribbean but the quality of the meat varies a great deal from species to species.

Turtles

A young green turtle, *Chelonia mydas*.

A hatchling leatherback.

Sea turtles feed around coasts and nest on beaches throughout the tropical seas of the world. They were once abundant, but their populations have been so reduced by indiscriminate hunting that strict conservation measures are required to save them from extinction.

Of the five species known in the Caribbean region, the green turtle, *Chelonia mydas*, is the most important economically. It is noted for its navigational ability, making migrations as far as 1,200 kilometres through open ocean to find the proper nesting area. The female green turtle lays her eggs above the high tide mark on carefully selected beaches. Approximately 100 eggs are laid at each nesting, with the female laying in an average of five nests each season at intervals of 10-15 days. The females do not lay each year, however, but in cycles of two, three or four years.

Once a nesting beach has been selected, the nesting process passes through a number of clearly defined stages during which the adult female may be on the beach for one or two hours. On arriving at the beach, the turtle crawls a little way up from the surf line and pushes her nose into the sand. This 'sand smelling' is repeated several times until the turtle is satisfied that a suitable beach sand is present, then she crawls up the remaining portion of the beach slope, pulling with her long front flippers and pushing with the shorter rear ones. On the flatter beach crest, where the sand is dry, she may wander around for quite a while until she selects the exact spot where nesting will take place. Digging begins in the chosen site with energetic swipes of the fore limbs and sand is thrown back over the body until a wide pit has been excavated. She settles down in one part of this body pit, firmly anchored by stretching out her flippers to the side, and commences digging the nest hole.

The eggs will be laid in a bottle-shaped hole about 60 to 80 centimetres deep which is excavated by alternate movements of the hind flippers. Each flipper digs out a small lump of sand and throws it away to one side. As the hole deepens, the sensitive tips of the flippers test the depth and carefully shape and smooth the walls until it is ready to receive the eggs. Oviposition, or egg-laying, then proceeds, the eggs falling into the nest hole two or three at a time over about half an hour.

The turtle next pulls sand into the hole with the hind flippers and packs it down firmly on top of the eggs. Once they are covered and the top of the hole obliterated, she starts working again with the front flippers. These now throw sand backwards to bury the whole

nest area and the turtle will move this way and that, throwing sand all the while, until it is almost impossible to detect the exact spot where nesting took place.

Her job completed, the female turtle must now return to the sea to feed until the next batch of eggs is ready for laying. During the covering up movements on the beach crest, the turtle has been re-orientating herself to the sea. She has been responding to visual signals which tell her that she must move towards the brightest part of the horizon. The sky over the nearby sea is brighter than that over the land, so she moves off in this direction. Her orientation to brightness is strengthened when she begins to move downhill on reaching the beach slope, and once she feels the surf on her body new vigour seems to be gained and she swims swiftly out to sea.

A few green turtles nest on some Caribbean islands but the vast majority of those that live and feed in the region nest in one of two important rookeries. Many migrate each year to the tiny, low-lying Aves Island to lay their eggs, while hundreds of others visit the Tortuguero Beach in Costa Rica. The relative seclusion and inaccessibility of these two areas accounts for the continued presence of nesting turtles which have been destroyed in so many other parts of the Caribbean. Active conservation programmes on Aves and in Costa Rica over the past few years have contributed greatly to the hatching success of the green turtle.

The green turtle lays her eggs in the warm sand and covers them before returning to the sea. The young turtles hatch out about 60 days later.

The turtle eggs which have been buried in the sand hatch in about 60 days and the hatchlings work their way to the surface by struggling and digging upwards together. Emergence from the nest occurs almost always at night and the baby turtles immediately scramble for the ocean. It is during this period that the hatchlings are most susceptible to predation by crabs. Even upon entering the sea, the baby turtles must still contend with sharks and other large fish as well as attacks by sea birds. Should emergence from the nest take place in daylight many of the hatchlings will be eaten by birds before they reach the sea.

It is still not known where the young sea turtles spend the first year of their life, although it is assumed that they swim far out to sea and feed on small floating crustaceans and molluscs. After about one year they return to inshore waters and feed primarily on the beds of underwater sea grasses and on the invertebrate animals encrusting the grass fronds.

It is estimated that, due to predation, only one or two turtles from each batch of about one hundred eggs laid survives to become sexually mature. Maturity is reached in from four to six years, when these survivors will join the other adults in their nesting migration and, thereby, complete the life cycle.

Man's Use of the Green Turtle

The past four centuries have witnessed a vast decline in the number of green turtles. Early sailing ships relied heavily upon this resource to provide meat during trips across the Atlantic, with the Cayman Islands playing an important role as a prime source of turtles. With increasing world population and the resulting demand for more protein, this type of exploitation has multiplied until today there are only remnants of the previously existing turtle populations, and many former nesting beaches are no longer visited by sea turtles.

There are three aspects of the life history of the green turtle which make it vulnerable to predation by man:

1 The female must leave the protection of the sea to lay her eggs.
2 Large numbers of turtles tend to lay at the same time.
3 The green turtle usually returns to the same beach several times during one nesting season.

Man has been responsible for the reduction of sea turtle populations by killing the adults at sea and while nesting on beaches and by taking large numbers of eggs from nesting beaches for food. In addition, the development of many former nesting beaches into resort areas has driven away the turtles.

Man does not use only green turtle eggs and meat for food, but makes turtle soup with the cartilage from under the shell. In addition to this 'calipee' from under the shell, the shiny plates from outside the carapace or back are used for jewellery and decorative work and small, whole shells are sold as tourist items. The skin, particularly from around the neck and shoulders, produces a good quality reptile leather. *Chelonia mydas* is one of the most valuable reptiles in the world and is reared for the market in some countries in special farms. The farm in Grand Cayman is one of the few in this region, but many Caribbean countries are protecting their sea turtle resources, especially the nests and the nesting females, in attempts to resuscitate the dwindling stocks.

The hawksbill turtle, *Eretmochelys imbricata*, produces the best quality turtle shell. This species feeds and nests in the same areas as the green turtle but is less common. Its food preference is for shellfish which are torn off rocks and crushed by the strong hawk-like beak before being swallowed whole. Like the green turtle, this species is often seen by divers as it sleeps underwater in holes and crevices on rocky shores and coral reefs.

The loggerhead turtle, *Caretta caretta*, which is slightly larger than either the green or hawksbill, can be recognised by its unusually large head and thick neck. In addition, loggerhead backs are often covered with encrusting barnacles, which are uncommon on other species. Loggerhead shell is little used in the turtle trade, although meat and eggs are taken readily.

The olive ridley turtle, *Lepidochelys olivacea*, is the smallest of the local sea turtle species and reaches only 100 centimetres maximum carapace length. It feeds largely on shellfish and other marine animal material collected in coastal waters. *Lepidochelys* nests two or three times each season at intervals of from 15-30 days on fine sand beaches. The batch of 90-140 eggs takes about 50 days to hatch. The olive ridley is not common in the region and is not recorded as nesting on any of the West Indian islands. It has been reported from the Caribbean coast of Colombia and also from Trinidad and the Guianas. In other parts of the world the ridley is exploited for its valuable skin.

The leatherback turtle, *Dermochelys coriacea*, is the largest of the world's sea turtle species and is quite different in general form from the others. It has very long fore flippers, as this reptile is adapted to aquatic

A hawksbill turtle, *Eretmochelys imbricata*, resting on a sandy beach.

Aggregate nesting of olive ridley turtles, *Lepidochelys olivacea*, on a beach in Suriname.

A leatherback turtle arriving in surf prior to climbing the beach for nesting.

life like the greens and hawksbills, but it does not have the same type of back or carapace. A tough skin supported by a mozaic of tiny bones covers the back and is produced into long hard ridges from head to tail. It is highly adapted for aquatic life, spending most of its life far out in the ocean, where it feeds on animal food, particularly jellyfish. It visits Caribbean beaches between March and September for nesting, and seems to prefer those with rough surf, coarse sand and a steep beach slope. A small number of leatherbacks nest on most West Indian islands but this species is more often seen in Trinidad, Colombia and the Guianas; very large nesting populations visiting Surinam and French Guiana each year. The nesting process is closely similar to that described for the green turtle and may occur as many as seven times each season at intervals of about 10 days. This species is the least popular with man, although in some parts of the region its eggs and meat are eaten.

Ecology and Conservation

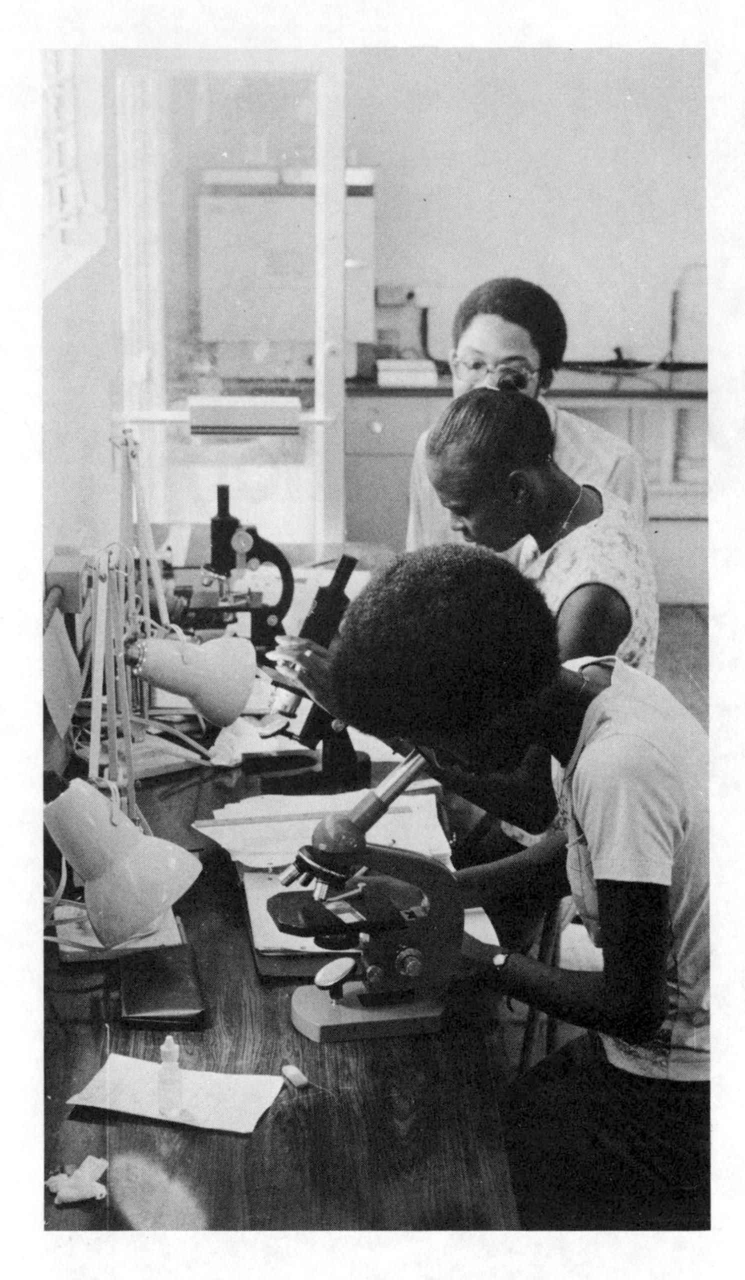

A young West Indian biologist at work in the laboratory. Her work here will complement and extend the knowledge gained in the environment itself.

We hear a lot today about **ecology**. What does the word mean? Essentially ecology is the study of the relationship of living things in their natural habitat. This includes both the interactions between the various organisms and the interactions between them and the physical world. The word comes from the Greek word 'oikos' meaning home or dwelling. Thus ecology is a study which must take place in the environment. Clearly laboratory studies and reading help in ecology but when all is said and done the ecologist must get out 'into the field'.

Even the simplest observation can be ecological; if you watch a humming bird feeding, collect shells on a beach, or watch a lizard catching flies you are being an ecologist in a simple way. To make these studies more worthwhile however, you should try and measure things, put numbers to your observations: How often does the humming bird come to that bougainvillia? How long does it feed at each flower? How many types of shells can you collect and how many of each type? How many flies does a lizard catch and how many lizards are there on the walls of your house?

Ecologists study living things in their environment in a variety of ways and with a variety of aims. Here are examples of some of the approaches:

1 Community ecologists are interested in one particular type of habitat, or a part of it. Thus such a person may be interested in the way in which the animals, plants and environment interact in a mangrove swamp or coral reef: What eats what? Who lives where? How many of each are there? Do the numbers change during the year? What happens if we change some important physical factor? (e.g. water level by drainage). Clearly, results of such studies are of vital importance to the conservationists as they tell him how much change a community can absorb before fundamental and perhaps irreversible changes take place. They also aid him in detecting alterations in the community and perhaps enable him to react quickly enough to prevent deleterious changes. Such studies are often called synecology.

2 Autecology is the study of one species in its habitat. Thus the autecologist may specialise in the ecology of the lobster for example. This worker will be concerned with the number of lobsters, their growth, the rate at which they produce eggs, what they eat, what eats them and so on. This work will take place in different communities as the adult lobster lives in the reef while the larva is planktonic (see p. 48). Studies of this type help us to understand and

possibly manage populations of various creatures. They are particularly important in the fishing industry (are the fishermen taking too many lobsters?) agriculture (what factors contribute to the high populations of pink boll worms in cotton?) and medicine (how can we best control populations of malaria-carrying mosquitoes?).

Modern ecology is particularly concerned with the two main aspects of synecology and autecology. Many synecologists are studying the way in which energy flows through a community. Virtually all of the energy in a community is derived from photosynthesis. Thus in a mangrove swamp the trees trap the sun's energy. Some of this is passed to animals that eat the trees. More of it is passed on when the leaves fall and rot on the mud. Here herbivores and detritus-eating creatures eat the leaves and their break-down products and then are themselves eaten by carnivores. There may be imports into, and exports from, the community; detritus washing in from the sea, growing fishes migrating out to the nearby reefs. The extent to which a community captures energy may depend on a variety of factors. For example, arctic environments may be limited by temperature. Similarly productivity in the open tropical sea is usually very low due to the reduced level of important nutrients such as nitrate and phosphate. Such studies can tell us fairly precisely what a community may produce in terms of energy in a given time and so to what extent it can be cropped or exploited.

Population studies on the other hand are an off-shoot of autecology and are concerned particularly with the factors that control the numbers of a given creature in the community. Generally speaking, ecologists recognise two major types of factors controlling population size, density independent and density dependent factors. The former do not depend on the size of the population. For example, a hurricane may cause wide spread mortality irrespective of the density of the population. However starvation through over exploitation of a food source will increase with rising population density. Some forms of epidemic are worse at high population densities. Large populations will also encourage the greater increase of predators and parasites. Such studies are of great importance in managing populations of commercially important species, either pests or food animals and plants. It also helps us to understand some of the problems that we humans face with our rapidly rising population. Problems such as the production of sufficient food, the control of disease and the management of natural resources.

Research is still needed if we are to fully understand our regional ecosystems.

Undergraduates in a biology class at the Barbados campus of the University of the West Indies.

Ecologists make use of a wide variety of techniques. Their work may involve a good deal of quiet observation and recording (of say feeding habits of lizards). He may also need to know the numbers and distribution of various numbers of the community. If the community is too large for him to record it all he will sample small parts of it and extrapolate from these to give an estimate of total numbers. He may well have to capture, count and measure animals of interest. Frequently he will mark or tag these animals so that he can recognise them again. Subsequent capture of tagged individuals may give him information about the growth, life span and mobility of the animal. The tags will vary according to the animal, of course. Thus insects may be chemically labelled or have a small blob of paint put on them. Fish often have labels fastened through their fins while birds will have a small metal ring placed round one leg. Mammals (mongooses, rats etc.) may have small pieces clipped from their ears or claws. Crustacea (crabs, lobsters) are difficult to tag as they moult their skins when they grow. However it is now possible to use tags which will remain with the animal even after a moult. It is important that tagged animals should not be hampered in any way by their tag as the ecologist is concerned that his marked animals should in every way be regarded as normal and not subject to greater mortality. Thus the bird ring must not be too heavy, the fish tag likely to attract predators or the lobster tag likely to hinder moulting.

If you find a tagged animal you should tell someone who can relay relevant information about it to the investigator. Sometimes the tag will request specific information e.g. date, time and place of capture or even the return of the whole animal to say, the Ministry of Agriculture. Please do all you can to assist the ecologist in these matters as the success of his experiments often depends on the co-operation of the public.

The ecologist may also have to obtain data on a variety of physical factors in his area of study; temperature, light levels, salinity, humidity, wind speed, rainfall and so on. Sometimes he will do this himself, sometimes he will request the information from governmental or regional institutions (e.g. weather stations).

There can be no doubt that enlightened conservation depends on reliable ecological data. Conservation after all is concerned with the maintenance of communities in their 'normal' state. Conservationists must know what is 'normal'; they must know what fluctuations are within the natural range and which are likely to be inevitable or deleterious. While perhaps it might be for the conservationists to define 'deleterious' it is for the ecologist to supply the hard data upon which such value judgements can be made.

As was said at the beginning of this section, ecological studies can be very simple. Anyone can contribute to our knowledge of the environment. Indeed around the world the amateur and part-time ecologists have been very active both in their acquisition of useful data and their valuable support of conservation. There are numerous examples of this type of contribution to the natural history and ecology of this area. Michael Humfrey's book on the shells of the Caribbean is the work of a Jamaican policeman; the Flora of Barbados, based in part on the flower collections of the Goodings; the publications, talks and display on Arawak and other archeological remains by Desmond Nicholson in Antigua, highlight three very different areas of interest. Perhaps it is unfair to pick just three names as throughout the Caribbean the part-timers have been and still are an influential and formative influence in ecology, natural history and conservation.

What can be done to preserve the wonders and beauties of the Caribbean Sea? Those readers who have

A group of students from the Barbados campus of the University of the West Indies, studying the distribution of sand dune plants.

read through to this point will already be aware of some of the problems that have arisen as a result of Man's impact on the marine habitats. Clearly this drift towards the lowering of the quality of our environment must be stopped and if possible reversed. Suitable legislation can, of course, help to do this but much more important are the attitudes and desires of the population, both resident and visiting. In this respect education is of vital importance both in the formal institutions of learning and outside among the adults, not least among those who are influential in our societies. One of the most valuable things that this education can do is to encourage a simple love and appreciation of the animals and plants around us. Once that sense of joy has been experienced the desire to preserve and protect that which is beautiful will follow automatically. This is one of the reasons why it is vital for teachers and educators to strive to get their pupils and students out into their environment as much as possible. Children

The key factor in a growing awareness of the need for conservation is education. The younger this education starts the better.

may dirty their clothes and hands but their minds will certainly not be sullied. No amount of chalk and talk can substitute for that experience.

Some of the advances in recent years have been double edged. Take for example scuba diving. This technique enables the trained diver to spend an hour or perhaps more under the sea at depths as great as 50 metres. The undersea world that is revealed to the diver is indeed a wonderful one but it is fragile and does not easily regenerate itself. Unscrupulous collecting and spear fishing can destroy the beauty of a reef in a few years. On the other hand, this technique together with the photographs that the diver brings back have been of immense benefit from the educational, scientific and aesthetic standpoint. Many other water sport activities which bring us more into contact with nature are also potentially harmful to the environment and great care must be taken to ensure that we derive the maximum benefits with the minimum of ill-effects.

Conservation as a general policy is not founded on a conservative desire to deny change for no good reason. Much more it is motivated by the need to preserve the goose that lays the golden eggs. Everyone in the Caribbean realises that a proper supply of protein is essential to our well being, and that various marine animals can contribute to that supply. The enlightened conservationist is interested not in reducing such exploitation but rather in maximising it, *by having a regard to the long term as well as the short term* consequences. As has been said already one cannot crop more from a group of animals than it produces without eventually bringing about its destruction; better reduce the exploitation now and know that the catches will still be the same in ten years time. It is in this vital area that the conservationist and the ecologist should come together to guide and advise those who crop the sea. But they must always be at pains to explain the logic of their advice for today's dollar speaks louder than tomorrow's in many minds. Again we find ourselves back to the basic need for education. This book has been dedicated to the youth of the Caribbean; but let not the older readers forget the responsibility they (and we) have for moulding the attitudes of those young people in whose hands the future of the Caribbean lies.

It has been said that the greatest measure of a people is the environment they leave behind for others. For the people of the Caribbean, the moment to choose the measure of our own greatness is here, now. Will we destroy our coral reefs, or preserve their richness? Will we fill in the mangrove swamps or allow their vital life processes to continue? Will we hunt and trap our wildlife to extinction, or help it to thrive? Will we continue to clutter our landscape, or keep it clean and beautiful? It is our decision, and it is we whom future generations will praise, or blame.

Glossary

The terms are defined here in a way appropriate to this book. Some of them may have a wider meaning than the definition given here.

Abyss The deepest parts of the oceans and seas where there is no light and thus no growing plants.

Adaptations The ways in which an animal or plant is suited to its particular way of life.

Alga (pl. **algae**) A group of plants which includes all the seaweeds. The term can be used to apply to a particular seaweed (e.g. an alga).

Aquaculture The farming of aquatic animals. (See page 52).

Aquatic Living in or pertaining to water, either fresh or salt.

Bacteria Microscopic unicellular organisms. Each cell is much less complex than any cell of an animal or plant. Although some cause disease, they are mostly involved in the breakdown of organic matter.

Barbel Tentacle-like growths from the lower jaw or mouth region of some fish. They are usually used as sensory organs for probing sand and mud.

Benthic Referring to the sea bed and the organisms found there.

Brittle stars See Echinoderms.

Carnivore A meat-eating animal.

Chlorophyll A coloured pigment in green plants essential for photosynthesis.

Cilium See Flagellum.

Cleaners, Cleaning Fish, Cleaning Shrimps. See page 63.

Commensalism See page 63.

Community An ecological term for any naturally occurring group of organisms living in the same environment.

Conch A group of snail-like molluscs. (See page 50).

Conservation The protection, management and wise use of all living and non-living, cultural and human resources.

Coral A general term applied to many coelenterate animals that form a horny or limy skeleton. The hard corals are those which are mainly responsible for reef formation. (See page 36).

Coral Reef A submarine structure mainly made of the accumulated skeletons of hard corals usually covered with living coral. Some reefs are dead and have no living coral, while others may have been raised above the sea level to form coral rock. (See page 36).

Crustaceans Members of the class Crustacea which is part of the phylum Anthropoda which also includes the insects. The crustaceans all have a jointed external skeleton and are mostly aquatic. The best known are the crabs and lobsters. (See page 46.)

Currents Movements of water created by either winds, tides or differences in salinity and/or temperature between water masses.

Cuticle A waxy layer secreted onto the surface of the leaf, one of whose functions is to reduce the evaporation from the leaf. (See page 14).

Detritus Small fragments of material having their origin from living creatures, either by breakdown or excretion.

Echinoderms Members of the phylum Echinodermata. There are five main groups (See page 44.)

Ecology See page 80.

Environment All the conditions or influences within a particular ecosystem which affect the organisms of that ecosystem.

Fauna All of the animals living in a particular place.

Filter Feeding See page 23.

Flagellium (pl. **flagella**) A tiny hair-like cell process used by microscopic animals and plants for locomotion (see page 68) Many larger animals have surfaces, such as gills, covered by many similar structures. In such large groupings they are termed cilia (sing. cilium).

Flora All the plants living in a particular place.

Food Chain and Food Web Sea grass is eaten by urchins which are eaten by fish. This is an example of a food chain. However, the situation is hardly ever as simple as that as other things may eat the sea grass, urchins will eat other plants and fish will eat food other than urchins. A food web attempts to describe these complex feeding interactions.

Gastropod The group of molluscs that includes snails and slugs.

Gills A filamentous structure used by marine animals for obtaining oxygen from the surrounding water. In some animals the gills may have other functions, e.g. food collection in filter feeding bivalves.

Habitat The specific, physical place where an organism lives e.g. in a hole, under a rock, on a reef.

Herbivore A plant eating animal.

Igneous Rocks Rocks which have at sometime been molten, for example, rocks derived from volcanic lava.

Immature Not old enough to breed.

Invertebrates Animals without backbones including worms, crustacea, echinoderms and molluscs among many other groups.

Isopod A member of the crustacean group, the Isopoda. Some of these creatures are ectoparasites of fish. (See page 62.)

Kilometre See metric measures.

Larva (pl. **larvae**) The juvenile stage of many animals. The larva usually is different in appearance from the adult and may lead a very different way of life.

Luminescence The light produced by animals and plants. Many creatures have the capability to do this including planktonic organisms, worms and fishes. On land the best known Caribbean example is the firefly.

Mammals Warm-blooded hairy vertebrates which develop in the womb and are suckled when young.

Mariculture See aquaculture.

Marine Living in or pertaining to the sea.

Metre See metric measures.

Metric Measures This book uses metric measures. Some useful equivalents are:—

1 centimetre is about 0.4 inch
30 centimetres are about 1 foot
1 metre is about 3.3 feet
1 kilometre is about 0.62 mile
1 kilogramme is about 2.2 pounds
1 litre is about 1.76 pints
1 hectare is about 2.5 acres

Molluscs Members of the phylum Mollusca. There are three main sub-groups: the snails and slugs which creep on a flat foot and often have a shell, the bivalves, relatively sedentary creatures with a double shell within which the animal lives, and the cephalopods which are active swimming predators including octopuses and squids. (See page 49.)

Mucus A slimy secretion containing protein often used by filter and suspension feeders for trapping food particles.

Mussel A bivalve mollusc.

Mutualism See page 63.

Nudibranchs Shell-less molluscs related to the snails; sometimes called sea-slugs.

Nutrients Substances which have food value and are necessary for healthy growth and development. Important nutrients for plants include nitrates and phosphates.

Oysters A group of bivalve molluscs members of which are often found growing on the roots of red mangrove trees. (See page 15.)

Parasite An animal or plant that lives on or in another animal or plant to the host's detriment. Ectoparasites live on the surface of their hosts whilst endoparasites live inside them.

Pelagic Belonging to the deep oceans

Pelvic and Pectoral Fins Most fish have two pairs of fins growing from their sides. The more forward and often slightly higher pair are the pectorals while the hinder and often lower pair are the pelvics.

Photosynthesis The manufacture of complex chemicals from carbon dioxide and water using light as the source of energy. This is usually a property of plants, the green pigment, chlorophyll being essential in the process.

Plankton The animals and plants which float in large bodies of water. The drifting plants constitute the phytoplankton while the animals make up the zooplankton. Planktonic creatures are most plentiful near the surface. (See also page 68.)

Polyp An individual of a colonial animal such as a coral. See page 36.

Population Members of the same species living in a community.

Predator A carnivorous animal. Its victim is called the prey.

Prey The victim of a carnivorous animal.

Prop-roots Roots growing out from stems, often tree trunks, at an angle which tends to support the plant. Red mangrove trees have many prop-roots. (See page 12).

Radula The file-like tongue of many snail-like molluscs, used for rasping their food.

Rectum The hindmost portion of the gut.

Salinity The saltiness of the sea. This varies from place to place. Ocean water (including that of the Caribbean) contains about 3.5 per cent. salt while in the Red Sea evaporation of the water may raise this to 4.0 per cent. In the Baltic the salinity is much less than this due to the inflow of river water diluting the concentration.

Scavenger An animal that feeds on dead or dying organisms.

School A group of fish.

Scuba Self-Contained Underwater Breathing Apparatus. A wide variety of equipment is now available to permit unattached divers to descend to moderate depths carrying compressed air in cylinders. Typical amateur divers can stay under water from 30 minutes to an hour.

Sea Anemones Simple animals belonging to the phylum Cnidaria which also includes the jellyfish. They are closely related to the corals.

Sea Cucumbers See Echinoderms.

Sea Grass See page 28.

Sea Squirts Sedentary filter feeding animals with a larva like a tadpole. Thought to be related to the vertebrates.

Sea Urchins See Echinoderms.

Sedimentary Rocks Rocks formed as sediments on the beds of the seas, rivers or lakes, e.g. sandstones and limestones.

Sedentary Applied to animals that cannot or do not move about much or at all, e.g. barnacles and sea anemones.

Spicules Tiny spines of hard material, usually calcium carbonate or silica, that support the tissues of an animal (e.g. sponges).

Starfish See Echinoderms.

Sponges Members of the phylum Porifora. These simple animals, although containing many cells, are of very simple organisation having no proper nervous system. They feed by sieving out food from the surrounding sea. (See page 43.)

Substrate The solid surface to which animals are attached or over which they walk. Some substrates (e.g. sand) may be burrowed into.

Suspension Feeding See page 23.

Symbiosis An intimate and mutually beneficial relationship between organisms of different species, e.g. the coral polyps and their zooxanthellae. (See page 64.)

Territory An area of the habitat usually occupied by a single animal of a particular species. The occupant will defend his territory against other members of his species. Territories may also be held by breeding pairs or by groups of the same species. (See page 58.)

Tide The periodic (twice per day) rise and fall of sea level which is caused by the gravitational pull of the sun and moon. The size of these tides depends on the phases of the moon.

Toxic Poisonous.

Turtles Members of a group of mainly aquatic reptiles which also includes the tortoises and terrapins.

Turtle Grass Marine flowering plants growing in shallow water where it may form beds of considerable size. (See page 28.)

Tunicates Sedentary filter feeding animals whose larva superficially resemble a tadpole and have many features which link them to the vertebrates.

Vertebrates Animals with backbones including the fish, amphibia, reptiles, birds and mammals.

Worms A term loosely used to describe long thin animals of worm-like appearance. (See page 45.)

Zonation See page 34.

Zooxanthellae The symbiotic algae living in the cells of some animals, especially corals.

Bibliography

Acker S. *Mangrove Ecology.* Oceans Magazine, Vol. 5 No. 4, 1972.

Böhlke J. E. and Chaplin C. C. G. *Fishes of the Bahamas and adjacent tropical waters.* Livingston Publishing Co. 1968.

Chaplin C. C. G. and Scott P. *Fishwatchers Guide.* Harrowood Books, 1972.

Darwin C. *Coral Reefs.* University of California Press, 1962.

Gore R. *Wild Nursery of the Mangroves.* National Geographic Vol. 151 No. 5, 1977.

Greenberg I. *Guide to Corals and Fishes of the Caribbean.* Seahawk Press, 1977.

Humfrey M. *Sea Shells of the West Indies.* Collins, 1975.

Morris P. A. *A Field Guide to Shells.* Houghton Mifflin Co., Boston, 1973.

Randall J. E. *Caribbean Reef Fishes.* T.F.H. Publications Inc. N.J., 1968.

Robas A. K. *South Florida's Mangrove.* Sea Grant Bulletin No. 4, 1970.

Sutty L. *Seasbell treasures of the Caribbean Reefs, The Bahamas and Bermuda* Took I. *Fishes of the Caribbean*

Walton Smith F. G. *Atlantic Reef Corals.* University of Miami Press, revised 1971.

Warmke G. L. and Abbott R. T. *Caribbean Seashells.* Dover Books, 1975.

Voss G. L. *Seashore Life of Florida and the Caribbean.* E. A. Seemann, 1976.